Georg F. Volkert

Tensor Fields on Orbits of Quantum States and Applications

Georg F. Volkert

Tensor Fields on Orbits of Quantum States and Applications

Südwestdeutscher Verlag für Hochschulschriften

Imprint

Any brand names and product names mentioned in this book are subject to trademark, brand or patent protection and are trademarks or registered trademarks of their respective holders. The use of brand names, product names, common names, trade names, product descriptions etc. even without a particular marking in this work is in no way to be construed to mean that such names may be regarded as unrestricted in respect of trademark and brand protection legislation and could thus be used by anyone.

Publisher:
Südwestdeutscher Verlag für Hochschulschriften
is a trademark of
Dodo Books Indian Ocean Ltd., member of the OmniScriptum S.R.L Publishing group
str. A.Russo 15, of. 61, Chisinau-2068, Republic of Moldova Europe
Printed at: see last page
ISBN: 978-3-8381-2053-9

Zugl. / Approved by: München, LMU, Diss., 2010

Copyright © Georg F. Volkert
Copyright © 2010 Dodo Books Indian Ocean Ltd., member of the OmniScriptum S.R.L Publishing group

Dissertation an der Fakultät
für Mathematik, Informatik und Statistik
der Ludwig-Maximilians-Universität München
vorgelegt im April 2010

Referent: Prof. Dr. Detlef Dürr
Korreferent: Prof. Dr. Giuseppe Marmo

Abstract

On classical Lie groups, which act by means of a unitary representation on finite dimensional Hilbert spaces $\mathcal{H}$, we identify two classes of tensor field constructions. First, as pull-back tensor fields of order two from modified Hermitian tensor fields, constructed on Hilbert spaces by means of the property of having the vertical distributions of the $\mathbb{C}_0$-principal bundle $\mathcal{H}_0 \to P(\mathcal{H})$ over the projective Hilbert space $P(\mathcal{H})$ in the kernel. And second, directly constructed on the Lie group, as left-invariant representation-dependent operator-valued tensor fields (LIROVTs) of arbitrary order being evaluated on a quantum state. Within the NP-hard problem of deciding whether a given state in a n-level bi-partite quantum system is entangled or separable (Gurvits, 2003), we show that both tensor field constructions admit a geometric approach to this problem, which evades the traditional ambiguity on defining metrical structures on the convex set of mixed states. In particular by considering manifolds associated to orbits passing through a selected state when acted upon by the local unitary group $U(n) \times U(n)$ of Schmidt coefficient decomposition inducing transformations, we find the following results: In the case of pure states we show that Schmidt-equivalence classes which are Lagrangian submanifolds define maximal entangled states. This implies a stronger statement as the one proposed by Bengtsson (2007). Moreover, Riemannian pull-back tensor fields split on orbits of separable states and provide a quantitative characterization of entanglement which recover the entanglement measure proposed by Schlienz and Mahler (1995). In the case of mixed states we highlight a relation between LIROVTs of order two and a class of computable separability criteria based on the Bloch-representation (de Vicente, 2007).

Zusammenfassung

Auf Klassischen Lie-Gruppen, die vermittels einer unitären Darstellung auf einem endlich dimensionalen Hilbert-Raum wirken, werden zwei Klassen von Tensorfeld-Konstruktionen betrachtet: Erstens, als Pull-back Tensorfelder zweiter Stufe, ausgehend von modifizierten Hermiteschen Tensorfeldern auf Hilbert-Räumen, welche mit der Eigenschaft ausgestattet sind, die vertikalen Distributionen auf dem $\mathbb{C}_0$-Prinzipal-Faser Bündel $\mathcal{H}_0 \to P(\mathcal{H})$ über dem projektiven Hilbert-Raum $P(\mathcal{H})$ im Kern zu besitzen. Und zweitens, vermittels einer direkten Konstruktion auf Lie-Gruppen, als links-invariante darstellungsabhängige Operator-wertige Tensorfelder beliebiger Stufe, die anhand eines Quantenzustands ausgewertet werden können. Im Rahmen des NP-harten Entscheidungsproblems, ob ein gegebener Zustand in einem aus zwei Komponenten zusammengesetzten Quantensystem verschränkt oder separabel ist (Gurvits, 2003), wird gezeigt, dass obige Konstruktionen eine geometrische Methode liefern, welche die traditionelle Vieldeutigkeit, metrische Strukturen auf der konvexen Menge gemischter Zustände zu definieren, umgeht. Insbesondere bezüglich Quantenzustands-Mannigfaltigkeiten, die durch Orbits der Schmidt-Dekomposition-induzierenden lokal unitären Gruppe $U(n) \times U(n)$ definiert sind, findet man folgende Resultate: Im Fall reiner Zustände wird gezeigt, dass Schmidt-Äquivalenz-Klassen, welche Lagrange Untermannigfaltigkeiten angehören, maximal verschränkte Zustände definieren. Dies impliziert eine stärkere Aussage, als diejenige, welche von Bengtsson formuliert wurde (2007). Zudem findet man, dass sich Riemannsche Pull-back Tensorfelder in zwei Anteile auf Orbits separabler Quantenzustands-Mannigfaltigkeiten zerlegen lassen und eine quantitative Beschreibung der Verschränkung liefern, die ein von Schlienz und Mahler vorgeschlagenes Verschränkungsmaß wiedergibt (1995). Im Fall gemischter Zustände schließlich, wird eine Relation zwischen Operator-wertigen Tensorfeldern und einer Klasse berechenbarer Separabilitäts-Kriterien, die auf der Bloch-Darstellung basieren (de Vicente, 2007), aufgezeigt.

Contents

	Preface	11
	Introduction	13

I Pull-back tensor fields   17

1 Tensor fields on Hilbert spaces   18
  1.1 Classical tensor fields on $\mathcal{H}_\mathbb{R}$ 19
  1.2 Hermitian tensor fields on $\mathcal{H}$ 22
  1.3 Tensor fields on $\mathcal{H}_0$ having the $\mathbb{C}_0$-generator in the kernel 23

2 The pull-back on orbits in $\mathcal{H}_0$ 28
  2.1 The general case 28
  2.2 Pull-back tensors induced by the defining representation 35
    2.2.1 U(2) in $\mathbb{C}^2$ 37
    2.2.2 U(3) in $\mathbb{C}^3$ 40
  2.3 Pull-back tensors induced by representations of unitary subgroups 46
    2.3.1 Two different representations of SU(2) on $\mathbb{C}^3$ 46
    2.3.2 Representations of U(1) on $\mathbb{C}^n$ 50
    2.3.3 Product representation of U(n)×U(n) on $\mathbb{C}^{n^2}$ 51

3 Application on the separability problem in composite systems $\mathcal{H}_A \otimes \mathcal{H}_B$ 55
  3.1 Traditional algebraic approach 55
  3.2 A geometric approach 60
  3.3 A distance function to separable states 66

II Operator-valued tensor fields   71

4 Operator-valued tensor fields on Lie groups   72
  4.1 Intrinsic defined tensor fields 72
  4.2 Representation-dependent tensor fields 73
  4.3 Sums of operator-valued tensor fields 77

5 Application on the separability problem in composite systems $\mathcal{D}(\mathcal{H}_A \otimes \mathcal{H}_B)$ 79
  5.1 The set of mixed quantum states 79
  5.2 The separability problem 82
    5.2.1 Approach in the Bloch-representation 82
    5.2.2 A connection to LIROVTs 84
    5.2.3 Example: Werner states for the case $n = 2$ 87

6 Conclusions and outlook 89
6.1 The relation between pull-back tensors and operator-valued tensors 89
6.2 The identification of entanglement measures 90
6.3 Dirac's program revisited: The role of tensors in the foundations of quantum mechanics 93

Appendices 97

A The GNS-construction in finite dimensions 97
A.1 Hilbert spaces from pure states 100
A.2 Hilbert spaces from mixed states 101

B Special morphisms: From embeddings to tensor products 102

References 106

Preface

The present work is the result of an international collaboration having its origin in the conference meeting *On the present status of quantum mechanics* held in Trieste in September 2005. Between a broad spectrum of alternative views, one of the speakers, Professor Giuseppe Marmo from the University of Naples Federico II, emphasized certain differential geometric aspects, being hidden in the usual 'orthodox' perspective of quantum mechanics.

Thus, the idea came up to investigate possible 'smooth' connections between hidden geometry and alternative views coming along non-local hidden variables theories in the mathematical foundations of quantum mechanics. This research idea has been gratefully subsidized by the DAAD (German Academic Exchange Service) and a partial result has been published under the title 'Classical tensors from quantum states' in [4]. Based on these results, a particular connection to quantum entanglement characterization was detected during the proceedings, providing one of the main topics of the present Ph.D. thesis.

With respect to this, I would like to thank following persons being in particular involved in this work: Professor Detlef Dürr from the University of Munich (LMU) who encouraged this work, for his teaching of clear perspectivizing, his confidence and his continual consulting during the realization of the project; Professor Giuseppe Marmo, for advising me through the topic of the thesis in the last two years and for teaching me consistently how to capture certain geodesics for solving mathematical problems on the infinite dimensional manifolds of 'finite dimensional manifolds'. Within the collaborating research group of Professor Marmo in Naples, I would like to express thanks to Doctor Paolo Aniello, for several discussions, bibliographic suggestions and his remarks on the performance of computations. Moreover, I thank Doctor Jesús Clemente-Gallardo from the University of Zaragoza, for prolific discussions and suggestions on *Mathematica* programming.

On the side of the Mathematisches Instituts of the Ludwig-Maximilians-Universität München I would like to thank also the *Arbeitsgruppe Bohmsche Mechanik*, especially my room-mate Doctor Sarah Römer, for various helpful remarks.

Furthermore, I would like to express my gratitude to the following persons: Doctor Victor Andrés Ferretti, for various motivating talks and the encouragement to re-parameterize critic (first) positively; graduate engineer Vittorio Ferretti for significant insights on 'complexity'; Florian Doering for generous stylistic emendations; Daniela Ibello for the networking regarding the substantial Neapolitan culture: They all participated in the process of this work.

Finally, I am grateful to my parents, graduate psychologist Giovanna Valli Volkert and graduate engineer Wolfgang Volkert, for their positive support, confidence and patience throughout my entire promotion period. This work is especially dedicated to them.

Georg F. Volkert
Munich, March 2010

Introduction

Tensor fields have played a particular role in the formulation of classical physical theories for a long time. The most prominent examples are related to Einstein's theory of gravitation and Maxwell's theory of electromagnetism. Certainly also on the footing of Newtonian mechanics and Boltzman's statistical mechanics, the macroscopic characterization of many particle systems, from fluids, solids and maximal coarse grained objects like the mechanical description of rigid bodies, we encounter tensor fields to be unavoidable to get into account the sophisticated non-linear aspects involved in the particular physical systems.

On the contrary, when we study the basic mathematical structures of a given *quantum* mechanical system, we hardly get involved with such tensorial notions. One of the reasons may rely on the fact that quantum theory is widely established by means of unitary transformations on a Hilbert space, where tensor fields would appear a priori 'without application'.

However, as it has been emphasized by Dirac in [30],

"Classical mechanics must be a limiting case of quantum mechanics.
We should thus expect to find that important concepts in classical mechanics correspond to important concepts in quantum mechanics and, from an understanding of the general nature of the analogy between classical and quantum mechanics, we may hope to get laws and theorems in quantum mechanics appearing as simple generalizations of well-known results in classical mechanics."

Hence, one should strongly expect to encounter also tensorial structures in quantum mechanics, whenever we deal with the problem of the quantum-classical transition to Newtonian physics. In the interplay with the developments and applications of unitary representation theory of classical Lie groups and associated Lie algebras, an identification of geometric structures in quantum mechanics has been proposed by several authors, starting with Dirac (with his introduction of quantum Poisson brackets [29, 30]), Weyl, Segal, Mackey and Strocchi [52–55, 73, 76, 79, 80]. A strict geometrical formulation of quantum mechanics is more recent and has been developed during the following years in [1, 7, 12–19, 21–24, 27, 38, 40, 41, 46, 59, 68, 70].

The notion of a manifold in quantum mechanics emerged in this regard very natural by means of the identification of homogeneous spaces being isomorphic to *orbits of quantum states* generated by unitary representations of some classical Lie groups acting on a fiducial quantum state. The general frame work on generating manifolds in this way has been commonly established as generalized coherent states method, being most prominent in the traditional questions of quantization and the quantum-classical limit associated to infinite dimensional Hilbert spaces [5, 65, 66]. The identification of tensor fields by means of Riemannian and symplectic structures on these manifolds has been discussed by Klauder in several papers [48–50].

In contrast, it became increasingly evident that the non-linear features of certain *submanifolds of quantum states* may encode some crucial additional properties about the corresponding quantum system which are *not* necessarily related to any classical limit, but nevertheless particularly useful in questions of quantum statistics (see e.g. [69,70,82]).

In this regard, we encounter submanifolds of quantum states also associated to *finite* dimensional Hilbert spaces, in particular when one deals with questions of entanglement classification in quantum information and quantum computation [10,51,75].

Even though these submanifolds appear at the first glance very similar to the construction of generalized coherent state manifolds in infinite dimensional Hilbert spaces, they differ essentially by taking into account representations which are not necessarily irreducible. In particular they involve tensor product representations and provide therefore an action generated by a *subgroup* of the full automorphism group of the given product Hilbert space. Here one observes a particular and very interesting link between Schmidt coefficients of singular value decomposed quantum states and the homogeneous space topology of orbits [51,75], motivating strongly the idea that a geometric characterization of the latter may come along with useful applications in the description of quantum systems in analogy to those examples found in the field of quantum statistics [69,70,82]. An explicit class of examples of Riemannian and pre-symplectic tensor fields on orbits of a bipartite entangled 2-level system has been computed in this regard in [9]. Notwithstanding these considerations, a general tensorial approach both to n-level systems of *arbitrary finite dimensions* and to *mixed* bi-partite entangled quantum states is still missing in the literature. As a matter of fact, geometric methods used for the identification of entangled and separable states are commonly associated in the literature with distance functions on the *whole convex space of mixed quantum states* [10], rather then on the above mentioned orbits of quantum states. Being not a manifold, but a stratified space with a border [40,41], this causes ambiguities and serious technical problems in the basic construction of tensor fields, which may be avoided by a restriction on subsets established by homogeneous spaces, i.e. smooth manifolds provided by orbits of quantum states.

The present work is based on two recent papers by Aniello, Marmo, Clemente-Gallardo and Volkert [3,4], where a tensorial approach on orbits of bi-partite pure states has been solidified with applications for arbitrary finite dimensions. It is the aim of the underlying work to expose the latter approach in detail including new, not published results within the generalized regime of bi-partite mixed states.

In particular there will be a consideration of two distinguished methods to identify tensor fields on Lie groups and related orbits of quantum states in dependence whether we deal with pure or mixed states. Correspondingly, the working program will be subdivided into two parts as follows:

The first part will essentially involve

1. an identification of tensor fields on Hilbert spaces, which encode the geometry of the space of pure quantum states,

2. the pull-back of these tensor fields to Lie groups and related orbits of vectors in the Hilbert space, which cover orbits of pure quantum states and

3. a classification of orbits of entangled pure quantum states by means of pulled-back tensor fields.

The second part will focus on

4. a construction of representation-dependent operator-valued tensor fields on Lie groups and their evaluation on quantum states, defining scalar valued tensor fields and

5. an application of these tensor fields on the separability problem for mixed quantum states.

These steps will be tackled in an inductive way by beginning with low-dimensional systems and by trying to discuss their constructive generalization to higher dimensional systems as far as possible.

Part I
Pull-back tensor fields

1 Tensor fields on Hilbert spaces

We will start in the Schrödinger-picture by setting the notion of a complex Hilbert space at the first place. Let us recall for this purpose the basic notion:

Definition 1.1 (Hilbert space). *A vector space V over the field $K = \mathbb{C}$ of complex numbers, endowed with an Hermitian inner product, i.e. a $\mathbb{C}$-sesquilinear map*

$$\langle \cdot \,|\, \cdot \rangle : V \times V \to \mathbb{C}, \tag{1.1}$$

is called complex Hilbert space

$$\mathcal{H} := (V, \langle \cdot \,|\, \cdot \rangle) \tag{1.2}$$

(in short: a Hilbert space $\mathcal{H}$), if it is Cauchy-complete in respect to the norm

$$\|\cdot\| := \langle \cdot \,|\, \cdot \rangle^{1/2} \tag{1.3}$$

defined by $\langle \cdot \,|\, \cdot \rangle$.

With this definition in the hand we note that Hilbert spaces may differ topologically from each other by the notion of dimension. Two classes are given in this regard by Hilbert spaces being finite dimensional and isomorphic to $\mathbb{C}^n$, and infinite dimensional and isomorphic to $l^2(\mathbb{N}, \mathbb{C})$.

To avoid technical demanding subtleties coming along the infinite dimensional case right here at the beginning, we will keep following convention:

Remark 1.2. *If not differently stated, we will restrict the following discussion to finite dimensional Hilbert spaces.*

In the spirit of Dirac's program of identifying classical appearing structures in quantum mechanics [29, 30], and based on a more recent survey on a geometrization program of quantum mechanics (see [23, 24]), it is the aim of this first chapter to review in details the existence of *canonically* constructed tensor fields on finite dimensional Hilbert spaces which arise as *classical* tensors, in the sense of classical Riemannian and classical symplectic differential geometry. For this purpose we will consider a decomposition

$$\langle \cdot \,|\, \cdot \rangle = \operatorname{Re} \langle \cdot \,|\, \cdot \rangle + i \operatorname{Im} \langle \cdot \,|\, \cdot \rangle. \tag{1.4}$$

of the available Hermitian inner product on a Hilbert space $\mathcal{H} \cong \mathbb{C}^n$ in a given basis $\{|e_j\rangle\}_{j \leq n}$, where we will find a constructive translation of the quadruple

$$(\{|e_j\rangle\}_{j \leq n}, \operatorname{Re} \langle \cdot \,|\, \cdot \rangle, \operatorname{Im} \langle \cdot \,|\, \cdot \rangle, i) \tag{1.5}$$

into a corresponding quadruple of tensor fields on the realification of the Hilbert space $\mathcal{H}_\mathbb{R} \cong \mathbb{R}^{2n}$. The translation goes as follows.

1.1 Classical tensor fields on $\mathcal{H}_\mathbb{R}$

Consider a Hilbert space $\mathcal{H}$ in a given basis $\{|e_j\rangle\}_{j\in J}$. By means of the Hermitian inner product $\langle \cdot | \cdot \rangle$, one finds then for a given vector $|\psi\rangle$ in $\mathcal{H}$ the expansion

$$|\psi\rangle = \sum_j^n \langle e_j |\psi\rangle |e_j\rangle := \sum_j^n z^j |e_j\rangle \qquad (1.6)$$

with dual basis vectors $\langle e_j |$ in $\mathcal{H}^*$. The Hermitian inner product $\langle \cdot | \cdot \rangle$ on $\mathcal{H}$ induces in this regard the bijection

$$\mathcal{H}^* \cong \mathrm{Lin}(\mathcal{H}, \mathbb{C}) \subset C^\omega(\mathcal{H}), \qquad (1.7)$$

relating the dual basis vectors to linear functionals

$$\langle e_j | \cdot \rangle : \mathcal{H} \to \mathbb{C}, \qquad (1.8)$$

which are in a subset of holomorphic functions on $\mathcal{H}$. The n $\mathbb{C}$-valued functionals imply 2n real valued 0-forms on $\mathcal{H}$ coming along the decomposition into a real and imaginary valued part

$$\langle e_j | \cdot \rangle := z^j(\cdot) = q^j(\cdot) + ip^j(\cdot), \qquad (1.9)$$

with real valued coordinate functions $q^j, p^j \in C^\infty(\mathcal{H}_\mathbb{R})$ on the realification $\mathcal{H}_\mathbb{R} \cong \mathbb{R}^{2n}$. These 0-forms yield then via the exterior derivative

$$d\langle e_j | \cdot \rangle = dz^j(\cdot) = dq^j(\cdot) + idp^j(\ ) \qquad (1.10)$$

a holonomic basis of 2n co-vector fields on $\mathcal{H}_\mathbb{R}$.

On the other hand, one observes that a given Hermitian inner product on $\mathcal{H}$ provides by definition a complex valued function on $\mathcal{H}^* \times \mathcal{H}$. The evaluation of

$$\langle \psi | \psi \rangle = \langle e_j | \bar{z}^j z^k | e_k \rangle = \sum_{j,k} \langle e_j | e_k \rangle \bar{z}^j z^k = \sum_{j,k} \delta_{jk} \bar{z}^j z^k = \sum_j \bar{z}^j z^j$$

defines in particular a real quadratic function

$$\langle \cdot | \cdot \rangle |_D = \sum_j \bar{z}^j(\cdot) z^j(\cdot) = \sum_j (q^j(\cdot))^2 + (p^j(\cdot))^2, \qquad (1.11)$$

on the diagonal $D \subset \mathcal{H}^* \times \mathcal{H}$. By identifying the diagonal in $\mathcal{H}^* \times \mathcal{H}$ with $D \equiv \mathcal{H}$ we set

$$\langle \cdot | \cdot \rangle |_\mathcal{H} \equiv \left(\sum_j \bar{z}^j z^j \right)(\cdot). \qquad (1.12)$$

To this quadratic function we associate an order-2 tensor field

$$(d|\cdot\rangle)^\dagger \otimes_F d|\cdot\rangle = \left(\sum_j d\bar{z}^j \otimes_F dz^j\right)(\cdot) \tag{1.13}$$

by considering a tensor-product $\otimes_F$ on the module $F = T_0^1(\mathcal{H}) \otimes T_1^0(\mathcal{H})$ of order 1 tensor fields on $\mathcal{H}$. The relation (1.13) follows by direct evaluation of the left hand side

$$\sum_{j,k} (dz^j(\cdot)|e_j\rangle)^\dagger \otimes_F dz^k(\cdot)|e_k\rangle = \sum_{j,k} \langle e_j| d\bar{z}^j(\cdot) \otimes_F dz^k(\cdot)|e_k\rangle$$

$$= \sum_{j,k} \langle e_j|e_k\rangle d\bar{z}^j(\cdot) \otimes_F dz^k(\cdot) = \sum_{j,k} \delta_{jk} d\bar{z}^j(\cdot) \otimes_F dz^k(\cdot).$$

Moreover, by means of the 2n 1-forms defined in (1.10), we find that a decomposition of (1.13) into a real and an imaginary part is given by[1]

$$\sum_j dq^j \otimes_F dq^j + dp^j \otimes_F dp^j + i(dq^j \otimes_F dp^j - dp^j \otimes_F dq^j)$$

$$= \sum_j dq^j \odot_F dq^j + dp^j \odot_F dp^j + i dq^j \wedge_F dp^j, \tag{1.14}$$

i.e. a Riemannian tensor field

$$G := \sum_j dq^j \odot_F dq^j + dp^j \odot_F dp^j \tag{1.15}$$

and a symplectic tensor field

$$\Omega := \sum_j dq^j \wedge_F dp^j. \tag{1.16}$$

By setting

$$J := G^{-1} \circ \Omega \tag{1.17}$$

we will find a (1,1)-tensor field [59]. The explicit evaluation of $G^{-1} \circ \Omega$ yields

$$J = \sum_{j,k} \left(\frac{\partial}{\partial q^j} \odot_F \frac{\partial}{\partial q^j} + \frac{\partial}{\partial p^j} \odot_F \frac{\partial}{\partial p^j}\right) \circ (dq^k \otimes_F dp^k - dp^k \otimes_F dq^k)$$

$$= \sum_{j,k} \delta_j^k dp^k \odot_F \frac{\partial}{\partial q^j} - \delta_j^k dq^k \odot_F \frac{\partial}{\partial p^j}$$

---

[1] We use here and in the following the convention:

$$\theta^j \odot \theta^k := \frac{1}{2}(\theta^j \otimes \theta^k + \theta^k \otimes \theta^j)$$

$$\theta^j \wedge \theta^k := \frac{1}{2}(\theta^j \otimes \theta^k - \theta^k \otimes \theta^j)$$

$$= \sum_{j,k} \delta_k^j dp^k \odot_F \frac{\partial}{\partial q^j} - \delta_k^j dq^k \odot_F \frac{\partial}{\partial p^j}$$

$$= \sum_j dp^j \odot_F \frac{\partial}{\partial q^j} - dq^j \odot_F \frac{\partial}{\partial p^j}. \qquad (1.18)$$

It follows that this (1,1)-tensor field has the property

$$J \circ J = -\mathbb{1}. \qquad (1.19)$$

Hence, it defines a complex structure on the realification $\mathcal{H}_\mathbb{R} \cong \mathbb{R}^{2n}$. The latter property can be shown by direct computation, where we find

$$\sum_{j,k}(dp^j \odot_F \frac{\partial}{\partial q^j} - dq^j \odot_F \frac{\partial}{\partial p^j}) \circ (dp^k \odot_F \frac{\partial}{\partial q^k} - dq^k \odot_F \frac{\partial}{\partial p^k})$$

$$= \sum_{j,k}(dp^j \odot_F \frac{\partial}{\partial q^j} - dq^j \odot_F \frac{\partial}{\partial p^j}) \circ (\frac{\partial}{\partial q^k} \odot_F dp^k - \frac{\partial}{\partial p^k} \odot_F dq^k)$$

$$= \sum_{j,k} -\delta_k^j \frac{\partial}{\partial q^j} \odot_F dq^k - \delta_k^j \frac{\partial}{\partial p^j} \odot_F dp^k$$

$$= \sum_{j,k} -\delta_k^j (\frac{\partial}{\partial q^j} \odot_F dq^k + \frac{\partial}{\partial p^j} \odot_F dp^k), \qquad (1.20)$$

implying

$$(J \circ J)_j^k = -\delta_j^k. \qquad (1.21)$$

In conclusion we have found a geometric translation of the algebraic structures coming along the quadruple

$$(\{|\epsilon_j\rangle\}_{j \leq n}, \operatorname{Re}\langle \cdot | \cdot \rangle, \operatorname{Im}\langle \cdot | \cdot \rangle, i) \qquad (1.22)$$

on a finite dimensional complex Hilbert space $\mathcal{H} \cong \mathbb{C}^n$ into a corresponding quadruple of tensor fields on the realification $\mathcal{H}_\mathbb{R} \cong \mathbb{R}^{2n}$,

$$(\{dx_j\}_{j \leq 2n}, G, \Omega, J) \qquad (1.23)$$

given by a holonomic basis of real valued 1-forms,

$$\{dx_j\}_{j \leq 2n} \qquad (1.24)$$

a Riemannian

$$G = \sum_{j=1}^{2n} dx^j \odot_F dx^j \qquad (1.25)$$

a symplectic

$$\Omega = \sum_{j=1}^n dx^j \wedge_F dx^{j+n} \qquad (1.26)$$

and a complex structure

$$J = \sum_{j=1}^{n} dx^{j+n} \odot_F \frac{\partial}{\partial x^j} - dx^j \odot_F \frac{\partial}{\partial x^{j+n}}. \tag{1.27}$$

Remark 1.3. *The triple* (G, Ω, J) *induces on* $\mathcal{H}_{\mathbb{R}}$ *a Kähler structure [23].*

1.2 Hermitian tensor fields on $\mathcal{H}$

Let V be a complex vector space isomorphic to a Hilbert space $\mathcal{H}$ in the category of abstract vector spaces. Then we may define a tensor field on $\mathcal{H}$ from the Hermitian inner product without using coordinates by considering a covariant tensor field

$$\tau_{\mathcal{H}}(X_v, X_w)(\psi) := \langle X_v(\psi) | X_w(\psi) \rangle, \tag{1.28}$$

defined by a contraction with vector fields within a trivialized tangent bundle

$$X_v : \mathcal{H} \to T\mathcal{H} \cong \mathcal{H} \times V \tag{1.29}$$

$$\psi \mapsto (\psi, X_v(\psi)) := (\psi, v). \tag{1.30}$$

Hence, we arrive to a constant *Hermitian* (0, 2)-tensor field

$$\tau_{\mathcal{H}}(X_v, X_w)(\psi) = \langle v | w \rangle, \tag{1.31}$$

i.e. it defines on each tangent space $T_\psi V$ over a point $\psi \in \mathcal{H}$ a Hilbert space and therefore an isomorphism

$$T_\psi V \cong \mathcal{H} \tag{1.32}$$

in the category of Hilbert spaces.

This tensor field can now be related to the constant classical tensor fields considered in the previous subsection by the following: According to (1.13)-(1.16) we have

$$G + i\Omega = \left(\sum_{j=1}^{n} d\bar{z}^j \otimes_F dz^j \right), \tag{1.33}$$

which suggests the notation

$$G + i\Omega \equiv \langle d\psi | \otimes_F | d\psi \rangle, \tag{1.34}$$

written compactly as

$$\langle d\psi | \otimes_F | d\psi \rangle := \langle d\psi \otimes_F d\psi \rangle. \tag{1.35}$$

Remark 1.4. *Here and in the following we intend with* $d |\psi\rangle \equiv |d\psi\rangle$ *the application of the exterior derivative on the coordinate functions* $z^j(\psi) = \langle e_j | \psi \rangle$ *by considering the Leibniz rule on point-wise products between scalar valued functions in* $Lin(\mathcal{H}) \subset \mathcal{F}(\mathcal{H})$, *i.e. for a given basis*

$\{|e_j\rangle\}_{j \in J}$ we set

$$d|\psi\rangle = |d\psi\rangle = \sum_{j=1}^{n} dz^j(\psi)|e_j\rangle \qquad (1.36)$$

with a 'non-moving frame'

$$d|e_j\rangle = 0 \quad \text{for all } |e_j\rangle \in \{|e_j\rangle\}_{j \in J}, \qquad (1.37)$$

unless the contrary is stated.

In this regard we find:

Proposition 1.5. $\langle d\psi| \otimes_F |d\psi\rangle$ is a constant Hermitian (0, 2,-tensor field on $\mathcal{H}$, i.e.

$$\tau_{\mathcal{H}} = \langle d\psi \otimes_F d\psi \rangle. \qquad (1.38)$$

Proof. In coordinates we have

$$(d|\cdot\rangle)^\dagger \otimes_F d|\cdot\rangle = \left(\sum_j d\bar{z}^j \otimes_F dz^j \right)(\cdot). \qquad (1.39)$$

Let $T\mathcal{H}$ the tangent bundle over $\mathcal{H}$ endowed according to (1.37) with a trivial connection. For sections $\mathcal{H} \to T\mathcal{H}$, denoted by $X_v : \psi \mapsto (\psi, v)$ and $X_w : \psi \mapsto (\psi, w)$, and their expansion

$$X_v = \sum_{j=1}^{n} v^j \frac{\partial}{\partial z^j} \qquad (1.40)$$

$$X_w = \sum_{j=1}^{n} w^j \frac{\partial}{\partial z^j} \qquad (1.41)$$

in a holonomic basis of $\mathcal{H}$, we find then a contraction

$$\left(\sum_{j=1}^{n} (dz^j)^\dagger(X_v) \otimes_F dz^j(X_w) \right)(\psi) = \sum_j \bar{v}^j w^j, \qquad (1.42)$$

implying

$$\langle d\psi \otimes_F d\psi \rangle (X_v, X_w)(\psi) = \langle v | w \rangle. \qquad (1.43)$$

□

1.3 Tensor fields on $\mathcal{H}_0$ having the $\mathbb{C}_0$-generator in the kernel

Geometrical structures on the space of pure quantum states may be 'viewed' on the associated Hilbert space by a particular modification of the Hermitian tensor field considered in the previous section in (1.38).

Let us recall for this purpose the space of pure quantum states. Motivated by the probabilistic interpretation of quantum mechanics it is defined by the set of equivalence classes $[\psi]$ of vectors in the Hilbert space which differ by a complex number according to the equivalence relation

$$|\psi_1\rangle \sim |\psi_2\rangle :\Leftrightarrow \exists \lambda \in \mathbb{C}_0 : |\psi_1\rangle = \lambda |\psi_2\rangle \tag{1.44}$$

on any two vectors $|\psi_1\rangle, |\psi_2\rangle \in \mathcal{H}_0$. The set of these equivalence classes establishes by definition the *projective Hilbert space* $\mathcal{P}(\mathcal{H})$, illustrated topologically by the identification of all points which lie on a complex ray with an excluded 'initial point' in the origin $\{0\} \in \mathcal{H}$ of the Hilbert space.

The projective Hilbert space may therefore be denoted also as the space $\mathcal{R}(\mathcal{H})$ of complex rays in $\mathcal{H}_0 := \mathcal{H} - \{0\}$. The punctured Hilbert space $\mathcal{H}_0$ on the other hand, becomes in this way the total space of a principal fiber bundle

$$\pi : \mathcal{H}_0 \xrightarrow{\mathbb{C}_0} \mathcal{R}(\mathcal{H}) \tag{1.45}$$

over $\mathcal{R}(\mathcal{H})$ with $\mathbb{C}_0$, as structure group being isomorphic to each fiber $\pi^{-1}([\psi])$ over a given base point $[\psi] \in \mathcal{R}(\mathcal{H})$.

Having a fiber bundle in the hand, we may introduce a connection 1-form θ to establish a decomposition of the tangent bundle $T\mathcal{H}_0$ over the total space $\mathcal{H}_0$ into a horizontal and a vertical sub-bundle. This decomposition is obtained in the following way.

At each point $|\psi\rangle$ we consider the space of orthogonal vectors

$$\{|\psi\rangle\}^\perp \equiv \{|\phi\rangle \in \mathcal{H}|\, \langle \phi|\psi\rangle = 0\}. \tag{1.46}$$

These vectors define the horizontal sub-bundle in $T_\psi V \cong \mathcal{H}$ according to (1.32). In simpler terms, the horizontal distribution at each point $|\psi\rangle$ is defined by the Kernel of $\langle \psi|d\psi\rangle$.

The connection 1-form on a principal fiber bundle is therefore defined by requiring

$$\theta(X) = 0 \Leftrightarrow: X \text{ is a horizontal vector field,} \tag{1.47}$$

and by the requirement to be the 'identity' on fundamental vector fields. For general vector fields they take values in the Lie-algebra of the structure group. In particular for the vertical vectors fields on $\mathcal{H}_0$ over $\mathcal{R}(\mathcal{H})$, generating the vertical foliation, we set

$$\theta(\Delta) = 1 \Leftrightarrow: \Delta \text{ is a generating vector field of dilatations,} \tag{1.48}$$

$$\theta(\Gamma) = i \Leftrightarrow: \Gamma \text{ is a generating vector field of U(1) transformations} \tag{1.49}$$

according to the decomposition of the structure group $\mathbb{C}_0$ into the real subgroups

$$\mathbb{C}_0 \equiv \mathbb{R}_0^+ \times U(1) \tag{1.50}$$

and its associated 1-dimensional real Lie sub-algebras. Vice versa, *given* the generating vertical vector fields on the realification $\mathcal{H}_\mathbb{R} - \{0\}$ according to

$$\Delta : \psi \mapsto (\psi, \psi), \tag{1.51}$$

$$\Gamma := J(\Delta) \tag{1.52}$$

written in coordinates as

$$\Delta = q_j \frac{\partial}{\partial q_j} + p_j \frac{\partial}{\partial p_j} \tag{1.53}$$

$$\Gamma = q_j \frac{\partial}{\partial p_j} - p_j \frac{\partial}{\partial q_j}, \tag{1.54}$$

we may recover the connection 1-form by verifying that

$$\theta = \frac{q_j dq_j + p_j dp_j + i(q_j dp_j - p_j dq_j)}{q_j^2 + p_j^2} \tag{1.55}$$

satisfies (1.48), (1.49). With these expressions in the hand, it is possible to reformulate the connection 1-form and the generating vertical vector fields on $\mathcal{H}_0$ by setting

$$\theta \equiv \frac{\langle \psi | d\psi \rangle}{\langle \psi | \psi \rangle} \tag{1.56}$$

$$\Delta + i\Gamma \equiv \left\langle \psi \left| \frac{\partial}{\partial \psi} \right. \right\rangle. \tag{1.57}$$

The latter follows from

$$\bar{z}_j \frac{\partial}{\partial \bar{z}_j} := (q_j - ip_j)\left(\frac{\partial}{\partial q_j} + i\frac{\partial}{\partial p_j}\right), \tag{1.58}$$

which defines a complex valued vector field acting on smooth complex valued functions defined on a real manifold – rather then acting on holomorphic resp. anti-holomorphic functions on a complex manifold.

Next we may find local embeddings of the base space into $\mathcal{H}_0$ by means of local sections

$$s_i : U_i \to U_i \times \mathbb{C}_0 \tag{1.59}$$

in a local bundle chart $U \times \mathbb{C}_0$ of $\mathcal{H}_0$ with $U_i \subset \mathcal{R}(\mathcal{H})$. Note that the image of s_i identifies a submanifold with vector fields which are *not* in the kernel of the connection 1-form θ. To 'see' the geometry of the projective Hilbert space $\mathcal{P}(\mathcal{H})$ on $\mathcal{H}_0$ we will need a tensor field on the total space $\mathcal{H}_0$ which is degenerate along vertical vector fields on $\mathcal{H}_0$. The previous Hermitian tensor field

$$\langle d\psi \otimes_F d\psi \rangle$$

in (1.38) is by definition non-degenerate and will therefore not be invariant under such an action indeed. A way out yields the following modified tensor field, which contains a conformal factor

$$\frac{\langle d\psi \otimes_F d\psi \rangle}{\langle \psi | \psi \rangle} \tag{1.60}$$

and essentially 'subtracts the vertical directions out' by means of the connection 1-form θ due to

$$\frac{\langle d\psi \otimes_F d\psi \rangle}{\langle \psi | \psi \rangle} - \theta \otimes_F \theta^\dagger. \tag{1.61}$$

In this way we find:

Proposition 1.6. *The covariant tensor (0,2)-tensor field*

$$\tau_{\mathcal{H}_0} := \frac{\langle d\psi \otimes_F d\psi \rangle}{\langle \psi | \psi \rangle} - \frac{\langle \psi | d\psi \rangle \otimes_F \langle d\psi | \psi \rangle}{\langle \psi | \psi \rangle^2} \tag{1.62}$$

is invariant under $\mathbb{C}_0$-transformations on $\mathcal{H}_0$ and contains the infinitesimal generator of $\mathbb{C}_0$ in its kernel.

Proof. The invariance of (1.62) under (point-independent) $\mathbb{C}_0$-transformations

$$|\psi\rangle \mapsto \lambda |\psi\rangle \tag{1.63}$$

follows for all $\lambda \in \mathbb{C}_0$ directly from the invariance of the Hermitian inner product under unitary transformations – including $U(1)$ – and the compensation of $\mathbb{R}_0^+$-dilatations within the conformal factor $\langle \psi | \psi \rangle^{-1}$.

The generating vector field of $\mathbb{C}_0$-transformations on $\mathcal{H}_0$ on the other hand is defined by

$$\left\langle \psi \left| \frac{\partial}{\partial \psi} \right. \right\rangle := q_j \frac{\partial}{\partial q_j} + p_j \frac{\partial}{\partial p_j} + i \left(q_j \frac{\partial}{\partial p_j} - p_j \frac{\partial}{\partial q_j} \right). \tag{1.64}$$

and implies

$$\tau_{\mathcal{H}_0} \left\langle \psi \left| \frac{\partial}{\partial \psi} \right. \right\rangle = \frac{\langle \psi | d\psi \rangle}{\langle \psi | \psi \rangle} - \frac{\langle \psi | d\psi \rangle \langle \psi | \psi \rangle}{\langle \psi | \psi \rangle^2}, \tag{1.65}$$

i.e.

$$\tau_{\mathcal{H}_0} \left\langle \psi \left| \frac{\partial}{\partial \psi} \right. \right\rangle = 0. \tag{1.66}$$

$\square$

In this way we conclude that

$$\tau_{\mathcal{H}_0} = \frac{\langle d\psi \otimes_F d\psi \rangle}{\langle \psi | \psi \rangle} - \frac{\langle \psi | d\psi \rangle \otimes_F \langle d\psi | \psi \rangle}{\langle \psi | \psi \rangle^2}$$

is a pulled-back tensor from the projective Hilbert $\mathcal{P}(\mathcal{H})$ space to $\mathcal{H}_0$, induced by the $\mathbb{C}_0$-fiber bundle submersion

$$\mathcal{H}_0 \to \mathcal{P}(\mathcal{H}). \tag{1.67}$$

Hence, with $\tau_{\mathcal{H}_0}$ we are able to relate a structure on $\mathcal{H}_0$ to a corresponding structure on the space of (pure) quantum states. Note in this regard that while a projective Hilbert space is known to admit a Kählerian structure [61], it is clear that $\tau_{\mathcal{H}_0}$ is neither a Kählerian nor a Hermitian tensor field, as long as it is defined as degenerate structure on the whole punctured Hilbert space $\mathcal{H}_0$, i.e. on the total space of the $\mathbb{C}_0$-bundle over the projective Hilbert space.

To illustrate how the geometry of $\mathcal{P}(\mathcal{H})$ may be 'viewed' on $\mathcal{H}$, we remark the following: For instance, if we consider a local section from $U \subset \mathbb{C}P^1 \cong S^2$ to $\mathcal{H}_0 \cong \mathbb{C}_0^2$, the section will look like a local part of S^2, say S^2 without a point. However, as the horizontal distribution is not involutive, it is not possible to identify S^2 as an 'integral leave' of the horizontal distribution. The best we can achieve is an orbit of $SU(2)$ in $\mathcal{H}_0 \cong \mathbb{C}_0^2$ which covers S^2, according to the Hopf fibration

$$S^3 \xrightarrow{S^1} S^2. \tag{1.68}$$

A geometric characterization finally, may then be achieved by a pull-back of $\tau_{\mathcal{H}_0}$ on $SU(2) \cong S^3$. We will discuss in more general terms such a pull-back procedure in the following chapter.

2 The pull-back on orbits in $\mathcal{H}_0$

2.1 The general case

As a next step we will work out a procedure how to apply the structure

$$\tau_{\mathcal{H}_0} = \frac{\langle d\psi \otimes_F d\psi \rangle}{\langle \psi | \psi \rangle} - \frac{\langle \psi | d\psi \rangle \otimes_F \langle d\psi | \psi \rangle}{\langle \psi | \psi \rangle^2}$$

for a tensorial characterization of manifolds being related to orbits of (pure) quantum states. As a preliminary step we will introduce the notion of an orbit in a Hilbert space, which we will mainly encounter in the following chapters.

Definition 2.1 ($\mathcal{G} \cdot_U |0\rangle$-orbit). *Let $\mathcal{G}$ be a finite dimensional Lie group and*

$$U : \mathcal{G} \to U(\mathcal{H}) \tag{2.1}$$

a unitary representation $U(\cdot)$ of elements $g \in \mathcal{G}$ in a Hilbert space $\mathcal{H}$. Then the subset

$$\mathcal{O} := \{|g\rangle \equiv U(g)|0\rangle \, | \, g \in \mathcal{G}\} \subset \mathcal{H} \tag{2.2}$$

is called $\mathcal{G} \cdot_U |0\rangle$-orbit of a fiducial vector $|0\rangle \in \mathcal{H}_0$.

Since quantum states are considered as points in the projective Hilbert space, which are complex rays in the Hilbert space, it may become clear that any $\mathcal{G} \cdot_U |0\rangle$-orbit is a covering of a corresponding orbit of quantum states. More specific, we have:

Proposition 2.2 (Orbit of pure quantum states). *For a given $\mathcal{G} \cdot_U |0\rangle$-orbit $\mathcal{O}$ there is an orbit of pure quantum states in the projective Hilbert space $\mathcal{R}(\mathcal{H})$ defined by a quotient space*

$$Q \cong \mathcal{O}/U(1). \tag{2.3}$$

Proof. In dependence of the fiducial vector $|0\rangle \in \mathcal{H}_0$ and the chosen representation U, one will have topological distinguished orbits, defined by the quotient space

$$\mathcal{O} \cong \mathcal{G}/\mathcal{G}_0 \tag{2.4}$$

with isotropy group

$$\mathcal{G}_0 := \{h \in \mathcal{G} | U(h) |0\rangle = |0\rangle\}. \tag{2.5}$$

Any given $\mathcal{G} \cdot_U |0\rangle$-orbit is now related to an *orbit of (pure) quantum states*

$$Q \cong \mathcal{G}/\mathcal{G}_0^{U(1)} \tag{2.6}$$

whenever we consider an extended subgroup defined by

$$\mathcal{G}_0^{U(1)} := \{h \in \mathcal{G} | U(h) |0\rangle = e^{i\alpha(h)} |0\rangle\}. \tag{2.7}$$

This implies the identification of points of a $\mathcal{G} \cdot_U |0\rangle$-orbit which differ by a $U(1)$-phase coming along the character representation of the subgroup $\mathcal{G}_0^{U(1)} \subset \mathcal{G}$. □

Remark 2.3. *Note that an orbit of quantum states is not equivalent to a system of generalized coherent states according to Perelomov [66], because we are not requiring completeness of our orbit of quantum states, or equivalently we are not requiring the representation to be irreducible.*

The following theorem provides an identification of tensorial structures on Lie groups $\mathcal{G}$ being induced by orbits of quantum states, or, more precise, their covering given by $\mathcal{G} \cdot_U |0\rangle$-orbits in a Hilbert space.

Theorem 2.4. *Let $\{\theta_j\}_{j \in J}$ be a basis of left-invariant 1-forms on $\mathcal{G}$ and let $\{X_j\}_{j \in J}$ be a Lie algebra basis of $\mathcal{G}$ with $\{iR(X_j)\}_{j \in J}$, its representation in the Lie-Algebra of $U(\mathcal{H})$ associated to a unitary representation $U : \mathcal{G} \to U(\mathcal{H})$. For a given $\mathcal{G} \cdot_U |0\rangle$-orbit $\mathcal{O}$ in $\mathcal{H}_0$ there exists a pull-back of the tensor field*

$$\tau_{\mathcal{H}_0} = \frac{\langle d\psi | \otimes_F | d\psi\rangle}{\langle \psi | \psi \rangle} - \frac{\langle \psi | d\psi\rangle \otimes_F \langle d\psi | \psi\rangle}{\langle \psi | \psi \rangle^2} \qquad (2.8)$$

from $\mathcal{H}_0$ to $\mathcal{G}$, defined by

$$\tau_{\mathcal{G}} := \left(\frac{\langle 0| R(X_j) R(X_k) |0\rangle}{\langle 0 | 0 \rangle} - \frac{\langle 0| R(X_j) |0\rangle \langle 0| R(X_k) |0\rangle}{\langle 0 | 0 \rangle^2} \right) \theta^j \otimes \theta^k. \qquad (2.9)$$

Proof. A $\mathcal{G} \cdot_U |0\rangle$-orbit $\mathcal{O}$ in $\mathcal{H}_0$ implies a family of actions defined by a representation homomorphism, which respects the differentiability class of the Lie group manifold, which is analytical for finite dimensional Lie groups [30].

This family of actions on the fiducial vector $|0\rangle \in \mathcal{H}_0$ implies therefore a smooth map

$$f_{\mathcal{G}} : \mathcal{G} \to \mathcal{O} \subset \mathcal{H}_0 \qquad (2.10)$$

$$g \mapsto U(g) |0\rangle := |g\rangle \qquad (2.11)$$

by means of the unitary representation U. This induces a pull-back of the coordinate functions

$$\langle e_j | \psi\rangle = z^j(\psi) \in \text{Lin}(\mathcal{H}) \subset \mathcal{F}(\mathcal{H}) \qquad (2.12)$$

on the Hilbert space to functions on the Lie group $\mathcal{G}$ according to

$$\langle \epsilon_j | g\rangle := f_{\mathcal{G}}^* z^j(\psi) \in \mathcal{F}(G). \qquad (2.13)$$

A vector valued 0-form

$$|\psi\rangle = \sum_j \langle e_j | \psi\rangle | e_j\rangle = \sum_j z^j(\psi) |e_j\rangle \qquad (2.14)$$

on the Hilbert space becomes in this way pulled-back to a vector valued 0-form

$$|g\rangle = \sum_j \langle e_j | g \rangle |e_j\rangle = \sum_j f_{\mathcal{G}}^* z^j(\psi) |e_j\rangle \tag{2.15}$$

on $\mathcal{G}$. Hence, we may set

$$f_{\mathcal{G}}^* |\psi\rangle := |g\rangle, \tag{2.16}$$

where the pull-back operation f^* acts on $|\psi\rangle$, similarly to the exterior derivative operation $d|\psi\rangle := |d\psi\rangle$ by acting on the coefficients $\langle e_j | \psi \rangle$ of the basis expansion. It follows then from

$$f^* \circ d = d \circ f^* \tag{2.17}$$

that

$$f_{\mathcal{G}}^* \tau_{\mathcal{H}_0} = f_{\mathcal{G}}^* \left(\frac{\langle d\psi \otimes_F d\psi \rangle}{\langle \psi | \psi \rangle} - \frac{\langle \psi | d\psi \rangle \otimes_F \langle d\psi | \psi \rangle}{\langle \psi | \psi \rangle^2} \right) \tag{2.18}$$

$$= \frac{\langle dg \otimes_{f^*F} dg \rangle}{\langle g | g \rangle} - \frac{\langle g | dg \rangle \otimes_{f^*F} \langle dg | g \rangle}{\langle g | g \rangle^2} \tag{2.19}$$

holds, where f^*F denotes the pull-back of the module $F := T_1^0(\mathcal{H})$ of scalar-valued 1-forms on $\mathcal{H}$ to scalar-valued 1-forms on the Lie group $\mathcal{G}$. To evaluate $d|g\rangle \equiv |dg\rangle$, we identify the left invariant 1-forms

$$U(g)^{-1} dU(g) \equiv iR(X_j)\theta^j, \tag{2.20}$$

on the Lie group manifold $\mathcal{G}$ with values in the operators $iR(X_j) \in u(\mathcal{H})$ of the Lie algebra representation. Here one finds

$$d|g\rangle = |dg\rangle = dU(g)|0\rangle = U(g)U(g)^{-1}dU(g)|0\rangle = iU(g)R(X_j)\theta^j|0\rangle \tag{2.21}$$

and therefore a pull-back tensor on the Lie group given by

$$\frac{\langle dg \otimes_{f^*F} dg \rangle}{\langle g | g \rangle} - \frac{\langle g | dg \rangle \otimes_{f^*F} \langle dg | g \rangle}{\langle g | g \rangle^2}$$

$$= \left(\frac{\langle 0 | R(X_j)R(X_k) | 0 \rangle}{\langle 0 | 0 \rangle} - \frac{\langle 0 | R(X_j) | 0 \rangle \langle 0 | R(X_k) | 0 \rangle}{\langle 0 | 0 \rangle^2} \right) \theta^j \otimes_{f^*F} \theta^k. \tag{2.22}$$

$\square$

We underline that the pull-back tensor construction which we have made explicit here depends on the choice of two basic 'ingredients', namely

- the fiducial quantum state $|0\rangle \in \mathcal{H}_0$, and
- the Lie algebra representations of the Lie group $\mathcal{G}$.

Once these two ingredients are chosen, the theorem above implies several immediate statements. They are listed as follows.

Corollary 2.5. $\tau_\mathcal{G}$ *decomposes into a symmetric and an anti-symmetric tensor field*

$$\tau_\mathcal{G} \equiv G + i\Omega, \tag{2.23}$$

on $\mathcal{G}$ according to

$$G = \left(\frac{\langle 0| [R(X_j), R(X_k)]_+ |0\rangle}{\langle 0|0\rangle} - \frac{\langle 0| R(X_j) |0\rangle \langle 0| R(X_k) |0\rangle}{\langle 0|0\rangle^2} \right) \theta^j \odot \theta^k. \tag{2.24}$$

$$\Omega = \left(\frac{\langle 0| [R(X_j), R(X_k)]_- |0\rangle}{\langle 0|0\rangle} \right) \theta^j \wedge \theta^k. \tag{2.25}$$

Proof. The identification of a real symmetric and an imaginary antisymmetric part of the pulled back tensor field is provided by means of the anti-commutators and commutators defined according to

$$[R(X_j), R(X_k)]_+ := \frac{1}{2}(R(X_j)R(X_k) + R(X_k)R(X_j)) \tag{2.26}$$

$$[R(X_j), R(X_k)]_- := \frac{1}{2i}(R(X_j)R(X_k) - R(X_k)R(X_j)). \tag{2.27}$$

□

Corollary 2.6. $\tau_\mathcal{G}$ *defines a pull-back tensor field on $\mathcal{G}$, induced by the projection*

$$\mathcal{G} \to \mathcal{G}/\mathcal{G}_0 \cong \mathcal{O}, \tag{2.28}$$

from a covariant tensor field on a $\mathcal{G} \cdot_U |0\rangle$-orbit $\mathcal{O}$.

Proof. $\mathcal{O}$ is a homogeneous space providing a smooth submanifold in $\mathcal{H}_0$. The restriction of $\tau_{\mathcal{H}_0}$ on this submanifold defines a covariant tensor field. The pull-back of this restricted tensor field on $\mathcal{G}$ is then induced by means of the action of the group $\mathcal{G}$,

$$f: \mathcal{G} \to \mathcal{O} \subset \mathcal{H}_0 \tag{2.29}$$

$$g \mapsto U(g)|0\rangle := |g\rangle. \tag{2.30}$$

□

Remark 2.7. *In more general terms, we may identify for a given projection $\mathcal{G} \to \mathcal{G}/\mathcal{G}_0$, the pull-back of the full covariant tensor algebra on $\mathcal{G}/\mathcal{G}_0$ to a sub-algebra of the covariant tensor algebra on $\mathcal{G}$.*

In this regard we find:

Corollary 2.8. $\tau_{\mathcal{G}}$ *defines a pull-back tensor field on* $\mathcal{G}$, *induced by the projection*

$$\mathcal{G} \to \mathcal{G}/\mathcal{G}_0^{U(1)} \cong Q, \tag{2.31}$$

form a covariant tensor field on a orbit $Q \cong \mathcal{O}/U(1)$ *of quantum states.*

Proof. $\tau_{\mathcal{H}_0}$ is a pulled-back tensor from the projective Hilbert $\mathcal{P}(\mathcal{H})$ space to $\mathcal{H}_0$, induced by the $\mathbb{C}_0$-principal bundle projection

$$\mathcal{H}_0 \to \mathcal{P}(\mathcal{H}). \tag{2.32}$$

This follows from the invariance along $\mathbb{C}_0$-transformations and the fact that the infinitesimal generators of this action belong to the kernel of $\tau_{\mathcal{H}_0}$. Hence, the $U(1)$-principal bundle projection

$$\mathcal{O} \to \mathcal{O}/U(1) \subset \mathcal{P}(\mathcal{H}), \tag{2.33}$$

induces a pull-back tensor field on $\mathcal{O}$, which coincides with the restriction of $\tau_{\mathcal{H}_0}$ on $\mathcal{O}$. With corollary 2.6 it follows the statement, once we identify the projection $\mathcal{G} \to Q$ with the composition of projections

$$\mathcal{G} \to \mathcal{G}/\mathcal{G}_0 \to \mathcal{G}/\mathcal{G}_0^{U(1)} \cong Q. \tag{2.34}$$

$\square$

Corollary 2.9. *The covariant rank-2 tensor field* τ_G *associates to all group elements* $g \in \mathcal{G}$ *a left-invariant map*

$$T_g \mathcal{G} \times T_g \mathcal{G} \times \mathcal{R}(\mathcal{H}) \to \mathbb{C}, \tag{2.35}$$

defining a pure quantum state-dependent bilinear form on its Lie algebra $Lie(\mathcal{G}) \cong T_e \mathcal{G}$, *which coincides with a covariance matrix evaluated on a rank-1 projector in* $u^*(\mathcal{H})$.

Proof. The left invariant 1-forms

$$\theta^j : \mathcal{G} \to T^*\mathcal{G} \cong \mathcal{G} \times T_e^*\mathcal{G} \tag{2.36}$$

define a trivialization of the cotangent bundle of the group manifold $\mathcal{G}$. Once the representation is fixed we may discuss the dependence of the pulled-back tensor

$$\tau_G \equiv T_{jk} \theta^j \otimes \theta^k \tag{2.37}$$

on the choice of fiducial vector $|0\rangle$, which is completely encoded in the coefficients, collected into a coefficient matrix

$$(T_{jk}) \in \mathbb{C}^{d \times d}, \quad d = \mathrm{Dim}_\mathbb{R}(\mathcal{G}) \tag{2.38}$$

with

$$T_{jk} = \frac{\langle 0| R(X_j) R(X_k) |0\rangle}{\langle 0|0\rangle} - \frac{\langle 0| R(X_j) |0\rangle \langle 0| R(X_k) |0\rangle}{\langle 0|0\rangle^2}. \tag{2.39}$$

In particular by relating the fiducial vector $|0\rangle$ on the Hilbert space to a (Hermitian inner product dependent) rank-1 projector

$$\rho_0 := \frac{|0\rangle\langle 0|}{\langle 0|0\rangle} \tag{2.40}$$

in the dual of the Lie-algebra $u^*(\mathcal{H})$, we see that T_{jk} becomes related to the coefficients of a covariance matrix evaluated on such a rank-1 projector ρ_0 according to

$$T_{jk}(\rho_0) \equiv \mathrm{Tr}(\rho_0 R(X_j)R(X_k)) - \mathrm{Tr}(\rho_0 R(X_j))\mathrm{Tr}(\rho_0 R(X_k)). \tag{2.41}$$

Hence, we have a state dependent bilinear form on the Lie-algebra of $\mathcal{G}$, provided by the map

$$(X_j, X_k) \mapsto T_{jk}(\rho_0), \tag{2.42}$$

once we take into account the bijection

$$[|0\rangle] \leftrightarrow \rho_0 \tag{2.43}$$

between the space of rays $\mathcal{R}(\mathcal{H})$ and the set of rank-1 projectors in $u^*(\mathcal{H})$. $\square$

In this regard we may observe by considering a Bloch-representation expansion

$$\rho_0 \equiv \sum_a r^a \sigma_a \in u^*(\mathcal{H}) \tag{2.44}$$

with real components $r^a \in \mathbb{R}$, and $\{\sigma_a\}_{a \in J}$ a trace-orthonormal basis on $u^*(\mathcal{H})$, a decomposition of the tensor-coefficients into terms which are linearly and non-linearly dependent in the real components $r^a \in \mathbb{R}$ of the pure state ρ_0

$$\sum_a r^a \mathrm{Tr}(\sigma_a R(X_j)R(X_k)) - \sum_{a,b} r^a r^b \mathrm{Tr}(\sigma_a R(X_j))\mathrm{Tr}(\sigma_b R(X_k)). \tag{2.45}$$

Since the non-linear terms will vanish for the antisymmetric part of the tensor, it is sufficient to identify the linear part

$$L := \left(\frac{\langle 0|[R(X_j), R(X_k)]_+|0\rangle}{\langle 0|0\rangle}\right) \theta^j \odot \theta^k \tag{2.46}$$

as part of the symmetric tensor

$$G = \left(L - \frac{\langle 0|R(X_j)|0\rangle\langle 0|R(X_k)|0\rangle}{\langle 0|0\rangle^2}\right) \theta^j \odot \theta^k. \tag{2.47}$$

The degeneracy of the pull-back tensors is encoded by the choice of the fiducial state, which determines the dimension of the isotropy group.

In this way we may encounter possible applications within the framework of quantum mechanics by giving a geometrical description of quantum states, once the latter are associated to fiducial states within our pull-back construction. For this purpose we will focus in the following sections on how to compute explicitly these tensor coefficient matrices on orbits generated by means of finite dimensional unitary subgroups in finite dimensional Hilbert spaces.

2.2 Pull-back tensors induced by the defining representation

In the simplest case we have $\mathcal{G} = \text{Aut}(\mathcal{H}) = U(n)$. Along with the corresponding defining representation of the n^2-dimensional real Lie algebra $u(n) \cong \mathbb{R}^{n^2}$ we set

$$R(i\sigma_j) = \sigma_j, \tag{2.48}$$

where $\{i\sigma_j\}_{0 \leq j \leq n^2-1}$ denotes an orthonormal basis of the Lie algebra given by the generalized Pauli-matrices (multiplied by the imaginary unit i). These matrices are (without the imaginary unit i) Hermitian and share the properties[2]

$$\sigma_0 = \mathbb{1} \tag{2.49}$$

$$\text{Tr}(\sigma_j) = 0, \text{ for } j > 0 \tag{2.50}$$

$$\text{Tr}(\sigma_j \sigma_k) = 2\delta_{jk} \tag{2.51}$$

$$[\sigma_j, \sigma_k]_- = c_{jkl}\sigma_l \tag{2.52}$$

$$[\sigma_j, \sigma_k]_+ = \frac{2}{n}\delta_{jk}\sigma_0 + d_{jkl}\sigma_l, \tag{2.53}$$

where c_{jkl} and d_{jkl} denote full anti-symmetric and symmetric structure constants of the Lie-algebra $u(n)$. Note that the orthonormalization (2.51) is implied in the other properties within the decomposition

$$\sigma_j \sigma_k = [\sigma_j, \sigma_k]_+ + i[\sigma_j, \sigma_k]_-. \tag{2.54}$$

A (degenerate) tensor field on $U(n)$ is then induced by an orbit

$$\mathcal{Q} \cong U(n)/\mathcal{G}_0^{U(1)} \cong \mathcal{O}/U(1) \tag{2.55}$$

of quantum states generated by the defining representation in $\mathbb{C}^n$. It is given according to Theorem 2.4 by

$$\tau_\mathcal{G} = \left(\frac{\langle 0| \sigma_j \sigma_k |0\rangle}{\langle 0 |0\rangle} - \frac{\langle 0| \sigma_j |0\rangle \langle 0| \sigma_k |0\rangle}{\langle 0|0\rangle^2} \right) \theta^j \otimes \theta^k, \tag{2.56}$$

---

[2] We use here and in the following the convention:

$$[A, B]_+ := \frac{1}{2}(AB + BA)$$

$$[A, B]_- := \frac{1}{2i}(AB - BA).$$

unveiling according to Corollary 2.5 a decomposition

$$\tau_\mathcal{G} = G + i\Omega \tag{2.57}$$

into a symmetric and anti-symmetric structure

$$G = \left(\frac{\langle 0 | [\sigma_j, \sigma_k]_+ | 0 \rangle}{\langle 0 | 0 \rangle} - \frac{\langle 0 | \sigma_j | 0 \rangle \langle 0 | \sigma_k | 0 \rangle}{\langle 0 | 0 \rangle^2} \right) \theta^j \odot \theta^k. \tag{2.58}$$

$$\Omega = \left(\frac{\langle 0 | [\sigma_j, \sigma_k]_- | 0 \rangle}{\langle 0 | 0 \rangle} \right) \theta^j \wedge \theta^k. \tag{2.59}$$

By means of (2.52), (2.53) we conclude:

Proposition 2.10. *Let $\{i\sigma_j\}_{0 \leq j \leq n^2-1}$ be a trace-orthonormal basis of the Lie-algebra $u(n)$ and $U(g) = g$ a defining representation of $\mathcal{G} = U(n)$ on a Hilbert space $(\mathbb{C}^n, \tau_{\mathcal{H}_0})$. For any given $U(n) \cdot_U |0\rangle$-orbit $\mathcal{O}$ one finds then a pull-back structure $\tau_{U(n)} = G + i\Omega$, decomposed into*

$$L = \left(2\delta_{jk} + d_{jkl} \frac{\langle 0 | \sigma_l | 0 \rangle}{\langle 0 | 0 \rangle} \right) \theta^j \odot \theta^k, \tag{2.60}$$

$$G = \left(L - \frac{\langle 0 | \sigma_j | 0 \rangle \langle 0 | \sigma_k | 0 \rangle}{\langle 0 | 0 \rangle^2} \right) \theta^j \odot \theta^k, \tag{2.61}$$

$$\Omega = \left(c_{jkl} \frac{\langle 0 | \sigma_l | 0 \rangle}{\langle 0 | 0 \rangle} \right) \theta^j \wedge \theta^k, \tag{2.62}$$

i.e., a real symmetric tensors G (containing the linear part L) and an imaginary anti-symmetric tensor Ω.

Remark 2.11. *All symmetrized and anti-symmetrized tensor coefficients may considered as measurable quantities, provided by the fact of being related to expectation values of observables in $u^*(n)$.*

In this regard it may be instructive to focus on the linear part

$$L = \left(\frac{\langle 0 | [\sigma_j, \sigma_k]_+ | 0 \rangle}{\langle 0 | 0 \rangle} \right) \theta^j \odot \theta^k \tag{2.63}$$

of the symmetric tensor. It has not the generator $\sigma_0 = \mathbb{1}$ of the action of the 1-dimensional unitary subgroup, being responsible for the multiplicative factorization into

$$U(n) = e^{i\phi} SU(n), \quad e^{i\phi} \in U(1) \tag{2.64}$$

in its kernel. In agreement with section 1.3, the symmetric pull-back tensor G will be degenerate in respect to this 1-dimensional subgroup action, since we have

$$G_{(0k)} = \frac{\langle 0|[\sigma_0, \sigma_k]_+|0\rangle}{\langle 0|0\rangle} - \frac{\langle 0|\sigma_0|0\rangle \langle 0|\sigma_k|0\rangle}{\langle 0|0\rangle^2} \qquad (2.65)$$

$$= L_{(0k)} - \frac{\langle 0|0\rangle \langle 0|\sigma_k|0\rangle}{\langle 0|0\rangle^2} = 0, \qquad (2.66)$$

with

$$L_{(0k)} = \frac{\langle 0|\sigma_k|0\rangle}{\langle 0|0\rangle}, \qquad (2.67)$$

by setting $j = 0$ and $0 \leq k \leq n^2 - 1$.

2.2.1 U(2) in $\mathbb{C}^2$

Let us illustrate the corresponding pull-back structures on the simplest non-trivial example given for the case $n = 2$. Here we note that all symmetric structure constants d_{jkl} vanish, i.e. to evaluate the pull-back tensor on the group $U(2)$ from related orbits in the defining representation, we have to consider according to Proposition 2.10

$$L = 2\mathcal{E}_{jk}\theta^j \odot \theta^k. \qquad (2.68)$$

$$G = \left(2\delta_{jk} - \frac{\langle 0|\sigma_j|0\rangle \langle 0|\sigma_k|0\rangle}{\langle 0|0\rangle^2}\right)\theta^j \odot \theta^k. \qquad (2.69)$$

$$\Omega = \left(c_{jkl}\frac{\langle 0|\sigma_l|0\rangle}{\langle 0|0\rangle}\right)\theta^j \wedge \theta^k \qquad (2.70)$$

with $\{i\sigma_j\}_{0 \leq j \leq 4}$, the standard Pauli-matrices

$$\sigma_1 = \begin{pmatrix} 0 & 1 \\ 1 & 0 \end{pmatrix}, \quad \sigma_2 = \begin{pmatrix} 0 & -i \\ i & 0 \end{pmatrix}, \quad \sigma_3 = \begin{pmatrix} 1 & 0 \\ 0 & -1 \end{pmatrix}$$

and $\sigma_0 = 1$. To give an explicit evaluation we identify a generic fiducial vector in the two-dimensional punctured Hilbert space $\mathcal{H}_0 \cong \mathbb{C}^2 - \{0\}$,

$$|0\rangle = c^1 \begin{pmatrix} 1 \\ 0 \end{pmatrix} + c^2 \begin{pmatrix} 0 \\ 1 \end{pmatrix}, \qquad (2.71)$$

in an eigenvector basis of σ_3 with complex coefficients $c^j \in \mathbb{C}$. Note that we will *not* impose any normalization condition on the coefficients. We may take into account a normalization afterwards, whenever we consider the map

$$|0\rangle \mapsto \rho_0 := \frac{|0\rangle\langle 0|}{\langle 0|0\rangle} \in u^*(2). \tag{2.72}$$

In particular, by decomposing the complex coefficients of the fiducial vector in polar coordinates, we set[3]

$$|0\rangle \equiv \begin{pmatrix} r_1 e^{i\beta_1} \\ r_2 e^{i\beta_2} \end{pmatrix}, \quad r_j \in \mathbb{R}^+, \beta_j \in [0, 2\pi) \tag{2.73}$$

and find the coefficient matrices of the pull-back tensor field on $U(2)$, having on all tangent spaces constant values via the left-invariant bi-linear forms on its Lie algebra $u(2) \cong \mathbb{R}^4$ according to

$$(L_{(jk)}) = \begin{pmatrix} 1 & \frac{2\cos(\beta_1-\beta_2)r_1r_2}{r_1^2+r_2^2} & \frac{2\sin(\beta_1-\beta_2)r_1r_2}{r_1^2+r_2^2} & \frac{2r_1^2}{r_1^2+r_2^2}-1 \\ \frac{2\cos(\beta_1-\beta_2)r_1r_2}{r_1^2+r_2^2} & 1 & 0 & 0 \\ \frac{2\sin(\beta_1-\beta_2)r_1r_2}{r_1^2+r_2^2} & 0 & 1 & 0 \\ \frac{2r_1^2}{r_1^2+r_2^2}-1 & 0 & 0 & 1 \end{pmatrix} \tag{2.74}$$

$$(G_{(jk)}) = \begin{pmatrix} 0 & 0 & 0 & 0 \\ 0 & \frac{r_1^4-2\cos(2(\beta_1-\beta_2))r_1^2r_2^2+r_2^4}{(r_1^2+r_2^2)^2} & -\frac{2\sin(2(\beta_1-\beta_2))r_1^2r_2^2}{(r_1^2+r_2^2)^2} & \frac{2\cos(\beta_1-\beta_2)r_1r_2(r_2^2-r_1^2)}{(r_1^2+r_2^2)^2} \\ 0 & -\frac{2\sin(2(\beta_1-\beta_2))r_1^2r_2^2}{(r_1^2+r_2^2)^2} & \frac{r_1^4+2\cos(2(\beta_1-\beta_2))r_1^2r_2^2+r_2^4}{(r_1^2+r_2^2)^2} & \frac{2\sin(\beta_1-\beta_2)r_1r_2(r_2^2-r_1^2)}{(r_1^2+r_2^2)^2} \\ 0 & \frac{2\cos(\beta_1-\beta_2)r_1r_2(r_2^2-r_1^2)}{(r_1^2+r_2^2)^2} & \frac{2\sin(\beta_1-\beta_2)r_1r_2(r_2^2-r_1^2)}{(r_1^2+r_2^2)^2} & \frac{4r_1^2r_2^2}{(r_1^2+r_2^2)^2} \end{pmatrix} \tag{2.75}$$

$$(\Omega_{[jk]}) = \begin{pmatrix} 0 & 0 & 0 & 0 \\ 0 & 0 & \frac{2r_1^2}{r_1^2+r_2^2}-1 & -\frac{2\sin(\beta_1-\beta_2)r_1r_2}{r_1^2+r_2^2} \\ 0 & 1-\frac{2r_1^2}{r_1^2+r_2^2} & 0 & \frac{2\cos(\beta_1-\beta_2)r_1r_2}{r_1^2+r_2^2} \\ 0 & \frac{2\sin(\beta_1-\beta_2)r_1r_2}{r_1^2+r_2^2} & -\frac{2\cos(\beta_1-\beta_2)r_1r_2}{r_1^2+r_2^2} & 0 \end{pmatrix}. \tag{2.76}$$

By comparing the tensor coefficients with the matrix elements of a pure density state $\rho_0 \in u^*(2)$ associated to the fiducial vector $|0\rangle$

$$\rho_0 := \frac{|0\rangle\langle 0|}{\langle 0|0\rangle} = \frac{1}{r_1^2+r_2^2} \begin{pmatrix} r_1^2 & e^{i\beta_2-i\beta_1}r_1r_2 \\ e^{i\beta_1-i\beta_2}r_1r_2 & r_2^2 \end{pmatrix}, \tag{2.77}$$

we see that the above pull-back tensor coefficients contain the information on the superposition composition of the state in dependence of the chosen basis of eigenvectors, resp. the chosen initial experimental setup (for instance the orientation of a Stern-Gerlach magnet).

---
[3]Note that these coordinates may become singular in any single 1-dimensional subspace.

In particular, 'off-diagonal' terms appearing with a phase difference like

$$e^{i(\beta_1-\beta_2)}r_1r_2, \tag{2.78}$$

are usually associated with *interference* phenomena in typical quantum mechanical experiments[4]. Moreover, we discover also new relational terms, like the difference of the amplitudes

$$r_2^2 - r_1^2, \tag{2.79}$$

given in the symmetric coefficients G_{13} and G_{23} in (2.75). In this regard we may focus on the 'collapsed' tensor, once we perform the pull-back with a fiducial vector which is an eigenvector to σ_3, say

$$|0\rangle \equiv \begin{pmatrix} 1 \\ 0 \end{pmatrix}. \tag{2.80}$$

Here we find

$$(L_{(jk)}) = \begin{pmatrix} 1 & 0 & 0 & 1 \\ 0 & 1 & 0 & 0 \\ 0 & 0 & 1 & 0 \\ 1 & 0 & 0 & 1 \end{pmatrix} \tag{2.81}$$

$$(G_{(jk)}) = \begin{pmatrix} 0 & 0 & 0 & 0 \\ 0 & 1 & 0 & 0 \\ 0 & 0 & 1 & 0 \\ 0 & 0 & 0 & 0 \end{pmatrix} \tag{2.82}$$

$$(\Omega_{[jk]}) = \begin{pmatrix} 0 & 0 & 0 & 0 \\ 0 & 0 & 1 & 0 \\ 0 & -1 & 0 & 0 \\ 0 & 0 & 0 & 0 \end{pmatrix}. \tag{2.83}$$

In this way we may compare both triples (2.74)-(2.76) and (2.81)-(2.83) of pull-back tensors and conclude that there exists different degrees of degeneracy on the Lie group manifold $U(2)$ *in dependence of their fiducial vector*. In particular, whereas we find always a non-degenerate symmetric structure L, the symmetric tensor field G may admit a higher degree of degeneracy in the case of an eigenvector in contrast to a super-position vector as one can see from the comparison between (2.82) and (2.75). This difference on the degree of degeneracy is physically denoted in terms of the deviation of the expectation value associated to the σ_3 Pauli-matrix given by

$$G_{33} = \frac{\langle 0|\sigma_3^2|0\rangle}{\langle 0|0\rangle} - \frac{\langle 0|\sigma_3|0\rangle^2}{\langle 0|0\rangle^2} = \frac{4r_1^2 r_2^2}{(r_1^2+r_2^2)^2} \tag{2.84}$$

---

[4]We remark in this regard, that such quadratic terms may also occur in interference phenomena associated to classical electromagnetism.

in (2.75), which vanishes according to (2.82) if the fiducial state is in an eigenstate of σ_3, resp. iff one of the amplitudes r_j is zero.

For the anti-symmetric part Ω in (2.76), we find that the different degrees of degeneracy on the 4-dimensional group manifold $U(2)$ is encoded up to the normalization exactly in the interference terms, recovered by the sum

$$\Omega_{13} + i\Omega_{23} = \frac{e^{i(\beta_1 - \beta_2)} r_1 r_2}{r_1^2 + r_2^2}. \tag{2.85}$$

According to the Corollaries 2.6 and 2.8 in the previous section, we may identify these tensor fields on $U(2)$, induced by the composition of projections

$$U(2) \to S^3 \to S^2, \tag{2.86}$$

as pulled-back tensor fields from a $U(2)_U|0\rangle$-orbit

$$\mathcal{O} \cong U(2)/U(1) \cong S^3 \tag{2.87}$$

diffeomorphic to a 3-sphere, resp. as pull-back tensor fields from the 2-dimensional orbit of quantum states

$$\mathcal{O}/U(1) \cong U(2)/U(1) \times U(1) \cong S^2 \tag{2.88}$$

diffeomorphic to a 2-sphere.

2.2.2 U(3) in $\mathbb{C}^3$

By going to higher dimensional examples we will expect much richer structures. In particular by going to one higher dimension, we will encounter the 9-dimensional Lie group U(3). A commonly used basis for its Lie algebra is given by the Gell-Mann matrices

$$\lambda_1 = \begin{pmatrix} 0 & 1 & 0 \\ 1 & 0 & 0 \\ 0 & 0 & 0 \end{pmatrix} \quad \lambda_2 = \begin{pmatrix} 0 & -i & 0 \\ i & 0 & 0 \\ 0 & 0 & 0 \end{pmatrix} \quad \lambda_3 = \begin{pmatrix} 1 & 0 & 0 \\ 0 & -1 & 0 \\ 0 & 0 & 0 \end{pmatrix} \tag{2.89}$$

$$\lambda_4 = \begin{pmatrix} 0 & 0 & 1 \\ 0 & 0 & 0 \\ 1 & 0 & 0 \end{pmatrix} \quad \lambda_5 = \begin{pmatrix} 0 & 0 & -i \\ 0 & 0 & 0 \\ i & 0 & 0 \end{pmatrix} \quad \lambda_6 = \begin{pmatrix} 0 & 0 & 0 \\ 0 & 0 & 1 \\ 0 & 1 & 0 \end{pmatrix} \tag{2.90}$$

$$\lambda_7 = \begin{pmatrix} 0 & 0 & 0 \\ 0 & 0 & -i \\ 0 & i & 0 \end{pmatrix} \quad \lambda_8 = \begin{pmatrix} \frac{1}{\sqrt{3}} & 0 & 0 \\ 0 & \frac{1}{\sqrt{3}} & 0 \\ 0 & 0 & \frac{-2}{\sqrt{3}} \end{pmatrix} \quad \lambda_0 = \begin{pmatrix} 1 & 0 & 0 \\ 0 & 1 & 0 \\ 0 & 0 & 1 \end{pmatrix}. \tag{2.91}$$

An important difference to the lower dimensional case of U(2) is that we will have to take into account non-trivial symmetric structure coefficients denoted in the following by d_{jkl}. The pull-back tensor fields on $U(3)$ become in this Lie algebra basis written as follows

$$L = \left(2\delta_{jk} + \tilde{c}_{jkl}\frac{\langle 0|\lambda_l|0\rangle}{\langle 0|0\rangle}\right)\theta^j \odot \theta^k. \tag{2.92}$$

$$G = \left(2\delta_{jk} + \tilde{a}_{jkl}\frac{\langle 0|\lambda_l|0\rangle}{\langle 0|0\rangle} - \frac{\langle 0|\lambda_j|0\rangle\langle 0|\lambda_k|0\rangle}{\langle 0|0\rangle^2}\right)\theta^j \odot \theta^k. \tag{2.93}$$

$$\Omega = \left(c_{jkl}\frac{\langle 0|\lambda_l|0\rangle}{\langle 0|0\rangle}\right)\theta^j \wedge \theta^k. \tag{2.94}$$

Let us consider in this regard a fiducial vector associated to a 3-level quantum state. By using again a corresponding polar coordinate decomposition

$$|0\rangle \equiv \begin{pmatrix} r_1 e^{i\beta_1} \\ r_2 e^{i\beta_2} \\ r_3 e^{i\beta_3} \end{pmatrix} \in \mathbb{C}^3, \quad r_j \in \mathbb{R}^+, \beta_j \in [0, 2\pi) \tag{2.95}$$

we find an associated density state

$$\rho_0 := \frac{|0\rangle\langle 0|}{\langle 0|0\rangle} = \frac{1}{r_1^2 + r_2^2 + r_3^2}\begin{pmatrix} r_1^2 & e^{i\beta_2 - i\beta_1}r_1r_2 & e^{i\beta_3 - i\beta_1}r_1r_3 \\ e^{i\beta_1 - i\beta_2}r_1r_2 & r_2^2 & e^{i\beta_3 - i\beta_2}r_2r_3 \\ e^{i\beta_1 - i\beta_3}r_1r_3 & e^{i\beta_2 - i\beta_3}r_2r_3 & r_3^2 \end{pmatrix}, \tag{2.96}$$

where several distinguished 'interference terms' emerge in the off-diagonal elements. In the previous example we have seen that the anti-symmetric pull-back tensor coefficients on $U(2)$ contain the information on interference phenomena associated to a superposition of two vectors. In this regard we may focus again on the anti-symmetric part, this time by performing the pull-back on $U(3)$ with fiducial vectors given by (2.95). The result on the following page in (2.100) defines a degenerate anti-symmetric bi-linear form on the Lie algebra $u(3) \cong \mathbb{R}^9$, where we used the short-hand notation

$$\beta_{ab} := \beta_a - \beta_b \tag{2.97}$$

$$\beta_{abcb} := \beta_{ab} - \beta_{cb} \tag{2.98}$$

for the individual phase-differences, and

$$r_{ab} := r_a r_b \tag{2.99}$$

for the amplitude multiplications within the interference terms. It is interesting to observe that also new types of relations emerge like the difference of the individual amplitudes $r_a - r_b$, a feature which has been only 'detected' in the symmetric part (2.75) of the pull-back tensor in the previous example on U(2). On the 9-dimensional manifold $U(3)$, we find in this regard an anti-symmetric bilinear form on $u(3) \cong \mathbb{R}^9$, which is illustrated on the next page in (2.100).

$$(\Omega_{(jk)}) = \underbrace{\begin{pmatrix}
0 & 0 & 0 & 0 & 0 & 0 & 0 & 0 \\
0 & 0 & 0 & 0 & 0 & 0 & 0 & 0 \\
\frac{r_2^2 - r_1^2}{r_1^2 + r_2^2 + r_3^2} & \frac{r_1^2 - r_2^2}{r_1^2 + r_2^2 + r_3^2} & 0 & \frac{\sin(\beta_{23})r_2 r_3}{r_1^2 + r_2^2 + r_3^2} & -\frac{\cos(\beta_{23})r_2 r_3}{r_1^2 + r_2^2 + r_3^2} & \frac{\sin(\beta_{13})r_1 r_3}{r_1^2 + r_2^2 + r_3^2} & -\frac{\cos(\beta_{13})r_1 r_3}{r_1^2 + r_2^2 + r_3^2} & 0 \\
\frac{2\sin(\beta_{12})r_1 r_2}{r_1^2 + r_2^2 + r_3^2} & \frac{2\cos(\beta_{12})r_1 r_2}{r_1^2 + r_2^2 + r_3^2} & -\frac{2\sin(\beta_{12})r_1 r_2}{r_1^2 + r_2^2 + r_3^2} & \frac{\cos(\beta_{23})r_2 r_3}{r_1^2 + r_2^2 + r_3^2} & \frac{\sin(\beta_{23})r_2 r_3}{r_1^2 + r_2^2 + r_3^2} & \frac{\cos(\beta_{13})r_1 r_3}{r_1^2 + r_2^2 + r_3^2} & \frac{\sin(\beta_{13})r_1 r_3}{r_1^2 + r_2^2 + r_3^2} & 0 \\
\frac{\sin(\beta_{23})r_2 r_3}{r_1^2 + r_2^2 + r_3^2} & \frac{2\cos(\beta_{12})r_1 r_2}{r_1^2 + r_2^2 + r_3^2} & \frac{2\cos(\beta_{12})r_1 r_2}{r_1^2 + r_2^2 + r_3^2} & \frac{r_3^2 - r_1^2}{r_1^2 + r_2^2 + r_3^2} & \frac{\cos(\beta_{13})r_1 r_3}{r_1^2 + r_2^2 + r_3^2} & \frac{\sin(\beta_{13})r_1 r_3}{r_1^2 + r_2^2 + r_3^2} & \frac{\cos(\beta_{12})r_1 r_2}{r_1^2 + r_2^2 + r_3^2} & -\frac{\cos(\beta_{13})r_1 r_3}{r_1^2 + r_2^2 + r_3^2} & -\frac{\sqrt{3}\sin(\beta_{13})r_1 r_3}{r_1^2 + r_2^2 + r_3^2} \\
\frac{\cos(\beta_{23})r_2 r_3}{r_1^2 + r_2^2 + r_3^2} & \frac{\sin(\beta_{23})r_2 r_3}{r_1^2 + r_2^2 + r_3^2} & \frac{\cos(\beta_{13})r_1 r_3}{r_1^2 + r_2^2 + r_3^2} & \frac{\sin(\beta_{12})r_1 r_2}{r_1^2 + r_2^2 + r_3^2} & \frac{\cos(\beta_{12})r_1 r_2}{r_1^2 + r_2^2 + r_3^2} & -\frac{\sin(\beta_{12})r_1 r_2}{r_1^2 + r_2^2 + r_3^2} & \frac{\sin(\beta_{12})r_1 r_2}{r_1^2 + r_2^2 + r_3^2} & \frac{\sqrt{3}\cos(\beta_{13})r_1 r_3}{r_1^2 + r_2^2 + r_3^2} \\
\frac{\sin(\beta_{13})r_1 r_3}{r_1^2 + r_2^2 + r_3^2} & \frac{\cos(\beta_{13})r_1 r_3}{r_1^2 + r_2^2 + r_3^2} & \frac{\sin(\beta_{23})r_2 r_3}{r_1^2 + r_2^2 + r_3^2} & \frac{\sqrt{3}\sin(\beta_{13})r_1 r_3}{r_1^2 + r_2^2 + r_3^2} & \frac{\sqrt{3}\cos(\beta_{23})r_2 r_3}{r_1^2 + r_2^2 + r_3^2} & \frac{\sqrt{3}\sin(\beta_{23})r_2 r_3}{r_1^2 + r_2^2 + r_3^2} & 0 \\
0 & 0 & 0 & 0 & 0 & 0 & 0 & 0
\end{pmatrix}}_{} \quad (2.100)$$

For the symmetric part of the pull-back tensor on $U(3)$ it is clear that a richer variety of degenerate structures emerge than on $U(2)$, once we consider the 'collapse' of the fiducial vector to either 1- or 2-dimensional subspaces of the Hilbert space.

For pure technical reasons coming along the illustration of the symmetric part of the pull-back tensor, we may consider in the following the pull-back on the subgroup $SU(3)$ defining a corresponding bi-linear form on the 8-dimensional Lie algebra and define furthermore a decomposition into sub-matrices,

$$(G_{jk})) := \begin{pmatrix} (G^A_{(jk)}) & (G^C_{jk}) \\ (G^C_{jk}) & (G^B_{(jk)}) \end{pmatrix} \tag{2.101}$$

with

$$(G^A_{(jk)}) \subset (G_{(jk)}) :\Leftrightarrow \quad 1 \leqslant j, k \leqslant 4 \tag{2.102}$$
$$(G^B_{(jk)}) \subset (G_{(jk)}) :\Leftrightarrow \quad 5 \leqslant j, k \leqslant 8 \tag{2.103}$$
$$(G^C_{jk}) \subset (G_{(jk)}) :\Leftrightarrow \quad 1 \leqslant j \leqslant 4;\ 5 \leqslant k \leqslant 8. \tag{2.104}$$

Like in the case of the anti-symmetric tensor, the vector field along the 'missing' λ_0 generator will be in the kernel of the associated symmetric bi-linear form. On the following pages one finds the results for these sub-matrices. The bi-linear form associated to the symmetric tensor L is illustrated on the whole Lie algebra of $U(3)$ and will be always non-degenerate like in the case of $U(2)$.

This is in agreement with the fact that L corresponds to the 'conformal' Riemannian tensor field on $\mathcal{H}_0 \cong \mathbb{C}^3 - \{0\}$ while G and Ω are associated via pull-back with tensor fields from the space of rays $\mathcal{R}(\mathbb{C}^3) \cong \mathbb{C}P^2$.

$$(G^A_{jk_l}) = \begin{Bmatrix} \dfrac{r_1^4+r_2^2+(r_1^2+r_3^2)r_3^2-2\cos(2\beta_{12})r_{12}^2}{(r_1^2+r_2^2+r_3^2)^2} & \dfrac{2\sin(2\beta_{12})r_{12}^2}{(r_1^2+r_2^2+r_3^2)^2} & \dfrac{\cos(\beta_{23})(r_2^2+r_3^2)-\cos(\beta_{23})+2\cos(\beta_{1312})r_1^2)r_{23}}{(r_1^2+r_2^2+r_3^2)^2} \\ -\dfrac{2\sin(2\beta_{12})r_{12}^2}{(r_1^2+r_2^2+r_3^2)^2} & \dfrac{r_1^4+r_3^4+(r_1^2+r_2^2)r_3^2+2\cos(2\beta_{12})r_{12}^2}{(r_1^2+r_2^2+r_3^2)^2} & -\dfrac{(2\sin(\beta_{1312})-\sin(\beta_{23}))r_1^2+\sin(\beta_{23})(r_2^2+r_3^2))r_{23}}{(r_1^2+r_2^2+r_3^2)^2} \\ \dfrac{2\cos(\beta_{12})(r_2^2-r_1^2)r_{12}}{(r_1^2+r_2^2+r_3^2)^2} & \dfrac{2\sin(\beta_{12})(r_2^2-r_1^2)r_{12}}{(r_1^2+r_2^2+r_3^2)^2} & \dfrac{\cos(\beta_{13})(-r_1^2+3r_2^2+r_3^2)r_{13}}{(r_1^2+r_2^2+r_3^2)^2} \\ -\dfrac{(\cos(\beta_{23})+2\cos(\beta_{1312}))r_1^2)r_{23}}{(r_1^2+r_2^2+r_3^2)^2} & \dfrac{\cos(\beta_{13})(-r_1^2+3r_2^2+r_3^2)r_{13}}{(r_1^2+r_2^2+r_3^2)^2} & \dfrac{r_1^4+(r_2^2+r_3^2)r_1^2+r_3^2(r_2^2+r_3^2)}{(r_1^2+r_2^2+r_3^2)^2} \end{Bmatrix}$$

(2.105)

$$(G^B_{jk_l}) = \begin{Bmatrix} \dfrac{r_1^4+r_2^2+2\cos(2\beta_{13})r_3^2)r_1^2+r_3^2(r_2^2+r_3^2)}{(r_1^2+r_2^2+r_3^2)^2} & \dfrac{2\cos(\beta_{12})(r_1^2+r_2^2)-(\sin(\beta_{12})+2\sin(\beta_{1323}))r_3^2)r_{12}}{(r_1^2+r_2^2+r_3^2)^2} & \dfrac{\sqrt{3}\sin(\beta_{13})(r_1^2+r_2^2-r_3^2)r_{13}}{(r_1^2+r_2^2+r_3^2)^2} \\ \dfrac{(\sin(\beta_{12})(r_1^2+r_2^2)+2\sin(\beta_{1323}))r_3^2)r_{12}}{(r_1^2+r_2^2+r_3^2)^2} & \dfrac{(\sin(\beta_{12})(r_1^2+r_2^2)-2\sin(\beta_{1323})r_3^2)r_{12}}{(r_1^2+r_2^2+r_3^2)^2} & \dfrac{r_2^4+r_3^4+2\cos(2\beta_{23})r_{23}^2+r_1^2(r_2^2+r_3^2)}{(r_1^2+r_2^2+r_3^2)^2} \\ \dfrac{(\cos(\beta_{12})(r_1^2+r_2^2)-2\cos(\beta_{1323})r_3^2)r_{12}}{(r_1^2+r_2^2+r_3^2)^2} & \dfrac{2\sin(2\beta_{23})r_{23}^2}{(r_1^2+r_2^2+r_3^2)^2} & \dfrac{\sqrt{3}\sin(\beta_{23})(r_1^2+r_2^2-r_3^2)r_{23}}{(r_1^2+r_2^2+r_3^2)^2} \\ \dfrac{\sqrt{3}\sin(\beta_{13})(r_1^2+r_2^2-r_3^2)r_{13}}{(r_1^2+r_2^2+r_3^2)^2} & -\dfrac{\sqrt{3}\cos(\beta_{23})(r_1^2+r_2^2-r_3^2)r_{23}}{(r_1^2+r_2^2+r_3^2)^2} & \dfrac{3(r_1^2+r_2^2)r_3^2}{(r_1^2+r_2^2+r_3^2)^2} \end{Bmatrix}$$

(2.106)

$$(G^C_{jk_l}) =$$

$$\begin{Bmatrix} \dfrac{(\sin(\beta_{23})(r_2^2+r_3^2)-(\sin(\beta_{1312})+2\sin(\beta_{1312}))r_1^2)r_{23}}{(r_1^2+r_2^2+r_3^2)^2} & \dfrac{((2\cos(\beta_{1312})-\cos(\beta_{23}))r_1^2+\cos(\beta_{23})(r_2^2+r_3^2))r_{23}}{(r_1^2+r_2^2+r_3^2)^2} & \dfrac{\sin(\beta_{13})(-r_1^2+3r_2^2+r_3^2)r_{13}}{(r_1^2+r_2^2+r_3^2)^2} & \dfrac{2\sin(2\beta_{13})r_{13}^2}{(r_1^2+r_2^2+r_3^2)^2} \\ \dfrac{(\cos(\beta_{13})r_1^2-(\sin(\beta_{13})+2\sin(\beta_{1232}))r_2^2+\sin(\beta_{13})r_3^2)r_{13}}{(r_1^2+r_2^2+r_3^2)^2} & \dfrac{\sin(\beta_{13})(-r_1^2+3r_2^2+r_3^2)r_{13}}{(r_1^2+r_2^2+r_3^2)^2} & \dfrac{\cos(\beta_{12})(r_1^2+r_2^2)-2\cos(\beta_{1323})r_3^2)r_{12}}{(r_1^2+r_2^2+r_3^2)^2} \\ -\dfrac{(\cos(\beta_{13})r_1^2-(\cos(\beta_{13})+2\sin(\beta_{1232}))r_2^2+\cos(\beta_{13})r_3^2)r_{13}}{(r_1^2+r_2^2+r_3^2)^2} & \dfrac{\cos(\beta_{23})(-3r_1^2+r_2^2-r_3^2)r_{23}}{(r_1^2+r_2^2+r_3^2)^2} & \dfrac{(\sin(\beta_{12})-2\sin(\beta_{1323}))r_3^2-\sin(\beta_{12})(r_1^2+r_2^2))r_{12}}{(r_1^2+r_2^2+r_3^2)^2} \\ \dfrac{2\sqrt{3}\cos(\beta_{12})r_3^2r_{12}}{(r_1^2+r_2^2+r_3^2)^2} & \dfrac{\sqrt{3}(r_1^2-r_2^2)r_3^2}{(r_1^2+r_2^2+r_3^2)^2} & \dfrac{\sqrt{3}\cos(\beta_{13})(r_1^2+r_2^2-r_3^2)r_{13}}{(r_1^2+r_2^2+r_3^2)^2} \end{Bmatrix}$$

(2.107)

$$(L_{(jk)}) = \begin{pmatrix}
1 & \frac{2\cos(\beta_{12})r_1 r_2}{r_1^2+r_2^2+r_3^2} & \frac{2\sin(\beta_{12})r_1 r_2}{r_1^2+r_2^2+r_3^2} & \frac{r_1^2-r_2^2}{r_1^2+r_2^2+r_3^2} & \frac{2\cos(\beta_{13})r_1 r_3}{r_1^2+r_2^2+r_3^2} & \frac{2\sin(\beta_{13})r_1 r_3}{r_1^2+r_2^2+r_3^2} & \frac{2\cos(\beta_{23})r_2 r_3}{r_1^2+r_2^2+r_3^2} & \frac{2\sin(\beta_{23})r_2 r_3}{r_1^2+r_2^2+r_3^2} & 1-\frac{3r_3^2}{r_1^2+r_2^2+r_3^2} \\[4pt]
\frac{2\cos(\beta_{12})r_1 r_2}{r_1^2+r_2^2+r_3^2} & \frac{r_1^2+r_2^2}{r_1^2+r_2^2+r_3^2} & 0 & 0 & \frac{2\cos(\beta_{13})r_1 r_3}{r_1^2+r_2^2+r_3^2} & \frac{2\sin(\beta_{13})r_1 r_3}{r_1^2+r_2^2+r_3^2} & \frac{2\cos(\beta_{23})r_2 r_3}{r_1^2+r_2^2+r_3^2} & \frac{2\sin(\beta_{23})r_2 r_3}{r_1^2+r_2^2+r_3^2} & \frac{\sqrt{3}\cos(\beta_{12})}{\sqrt{3}(r_1^2+r_2^2+r_3^2)} \cdot 2 \\[4pt]
\frac{2\sin(\beta_{12})r_1 r_2}{r_1^2+r_2^2+r_3^2} & 0 & \frac{r_1^2+r_2^2}{r_1^2+r_2^2+r_3^2} & 0 & \frac{\sin(\beta_{13})r_1 r_3}{r_1^2+r_2^2+r_3^2} & -\frac{\cos(\beta_{13})r_1 r_3}{r_1^2+r_2^2+r_3^2} & \frac{\sin(\beta_{23})r_2 r_3}{r_1^2+r_2^2+r_3^2} & -\frac{\cos(\beta_{23})r_2 r_3}{r_1^2+r_2^2+r_3^2} & \frac{2\sin(\beta_{12})}{\sqrt{3}(r_1^2+r_2^2+r_3^2)} \\[4pt]
\frac{r_1^2-r_2^2}{r_1^2+r_2^2+r_3^2} & 0 & 0 & 1 & \frac{\cos(\beta_{13})r_1 r_3}{r_1^2+r_2^2+r_3^2} & \frac{\sin(\beta_{13})r_1 r_3}{r_1^2+r_2^2+r_3^2} & -\frac{\cos(\beta_{23})r_2 r_3}{r_1^2+r_2^2+r_3^2} & -\frac{\sin(\beta_{23})r_2 r_3}{r_1^2+r_2^2+r_3^2} & \frac{r_1^2-r_2^2}{\sqrt{3}(r_1^2+r_2^2+r_3^2)} \\[4pt]
\frac{2\cos(\beta_{13})r_1 r_3}{r_1^2+r_2^2+r_3^2} & \frac{\cos(\beta_{23})r_2 r_3}{r_1^2+r_2^2+r_3^2} & \frac{\sin(\beta_{23})r_2 r_3}{r_1^2+r_2^2+r_3^2} & \frac{\cos(\beta_{13})r_1 r_3}{r_1^2+r_2^2+r_3^2} & 1-\frac{r_2^2}{r_1^2+r_2^2+r_3^2} & 0 & \frac{\cos(\beta_{12})r_1 r_2}{r_1^2+r_2^2+r_3^2} & \frac{\sin(\beta_{12})r_1 r_2}{r_1^2+r_2^2+r_3^2} & -\frac{\sqrt{3}(r_1^2+r_2^2+r_3^2)}{\sqrt{3}(r_1^2+r_2^2+r_3^2)} \\[4pt]
\frac{2\sin(\beta_{13})r_1 r_3}{r_1^2+r_2^2+r_3^2} & \frac{\sin(\beta_{23})r_2 r_3}{r_1^2+r_2^2+r_3^2} & -\frac{\cos(\beta_{23})r_2 r_3}{r_1^2+r_2^2+r_3^2} & \frac{\sin(\beta_{13})r_1 r_3}{r_1^2+r_2^2+r_3^2} & 0 & 1-\frac{r_2^2}{r_1^2+r_2^2+r_3^2} & -\frac{\sin(\beta_{12})r_1 r_2}{r_1^2+r_2^2+r_3^2} & \frac{\cos(\beta_{12})r_1 r_2}{r_1^2+r_2^2+r_3^2} & -\frac{\sin(\beta_{13})r_1 r_3}{\sqrt{3}(r_1^2+r_2^2+r_3^2)} \\[4pt]
\frac{2\cos(\beta_{23})r_2 r_3}{r_1^2+r_2^2+r_3^2} & \frac{\sin(\beta_{23})r_2 r_3}{r_1^2+r_2^2+r_3^2} & \frac{\cos(\beta_{13})r_1 r_3}{r_1^2+r_2^2+r_3^2} & \frac{\sin(\beta_{13})r_1 r_3}{r_1^2+r_2^2+r_3^2} & \frac{\cos(\beta_{12})r_1 r_2}{r_1^2+r_2^2+r_3^2} & \frac{\sin(\beta_{12})r_1 r_2}{r_1^2+r_2^2+r_3^2} & 1-\frac{r_1^2}{r_1^2+r_2^2+r_3^2} & 0 & -\frac{\cos(\beta_{23})r_2 r_3}{\sqrt{3}(r_1^2+r_2^2+r_3^2)} \\[4pt]
\frac{2\sin(\beta_{23})r_2 r_3}{r_1^2+r_2^2+r_3^2} & \frac{\cos(\beta_{13})r_1 r_3}{r_1^2+r_2^2+r_3^2} & \frac{\sin(\beta_{13})r_1 r_3}{r_1^2+r_2^2+r_3^2} & \frac{\sin(\beta_{23})r_2 r_3}{r_1^2+r_2^2+r_3^2} & \frac{\sin(\beta_{12})r_1 r_2}{r_1^2+r_2^2+r_3^2} & \frac{\cos(\beta_{12})r_1 r_2}{r_1^2+r_2^2+r_3^2} & 0 & 1-\frac{r_1^2}{r_1^2+r_2^2+r_3^2} & \frac{\sin(\beta_{23})r_2 r_3}{\sqrt{3}(r_1^2+r_2^2+r_3^2)} \\[4pt]
1-\frac{3r_3^2}{r_1^2+r_2^2+r_3^2} & \frac{2\cos(\beta_{12})r_1 r_2}{\sqrt{3}(r_1^2+r_2^2+r_3^2)} & \frac{2\sin(\beta_{12})r_1 r_2}{\sqrt{3}(r_1^2+r_2^2+r_3^2)} & \frac{r_1^2-r_2^2}{\sqrt{3}(r_1^2+r_2^2+r_3^2)} & -\frac{\cos(\beta_{13})r_1 r_3}{\sqrt{3}(r_1^2+r_2^2+r_3^2)} & -\frac{\sin(\beta_{13})r_1 r_3}{\sqrt{3}(r_1^2+r_2^2+r_3^2)} & -\frac{\cos(\beta_{23})r_2 r_3}{\sqrt{3}(r_1^2+r_2^2+r_3^2)} & -\frac{\sin(\beta_{23})r_2 r_3}{\sqrt{3}(r_1^2+r_2^2+r_3^2)} & \frac{r_1^2+r_2^2+4r_3^2}{3(r_1^2+r_2^2+r_3^2)}
\end{pmatrix} \quad (2.108)$$

2.3 Pull-back tensors induced by representations of unitary subgroups

We have seen in the previous section how the choice of a fiducial vector selects a bi-linear form on a Lie algebra, which has been associated to a pull-back tensor field from an orbit to the corresponding Lie group. In this section we will encounter the fact that even though we fix the choice of the fiducial vector, we will still have a variety of possibilities to realize different pull-back tensor fields on the Lie group and therefore different bi-linear forms on the Lie algebra: In particular one has more than one representation of the given Lie group $\mathcal{G}$ for considering an induced pull-back construction. Of course a general treatment would imply to encounter a comprehensive representation theoretical framework which would run out of the scope of this underlying work. Instead of making general claims, we may nevertheless discuss the difference between different representations in terms of our here presented pull-back procedure from a given Hilbert space in finite dimensions by focusing on special classes of examples.

2.3.1 Two different representations of SU(2) on $\mathbb{C}^3$

The simplest non-trivial example is given by

$$\mathcal{G} \equiv SU(2), \tag{2.109}$$

acting on a 3-dimensional Hilbert space, inducing the representation dependent map

$$f^U : SU(2) \rightarrow \mathbb{C}^3 \tag{2.110}$$

$$g \mapsto U(g)\ket{0} := \ket{g}. \tag{2.111}$$

which is generated by the representation dependent Lie algebra-valued 1-forms

$$U(g)^{-1}dU(g) = iR(i\sigma^j)\theta_j(g), \tag{2.112}$$

according to the Lie algebra- and associated Lie group representations

$$\begin{array}{ccc} SU(2) & \xrightarrow{U} & U(3) \\ \exp \uparrow & & \exp \uparrow \\ su(2) & \xrightarrow{iR} & u(3), \end{array}$$

with the Pauli-matrices $\sigma_j \in su^*(2)$.

In particular, we would like to see the difference between a reducible and an irreducible representation. For this purpose we note that the Lie-algebra generators $i\lambda_j$ of $SU(3)$ given by the Gell-Mann-matrices, will establish a reducible representation of $SU(2)$ on $\mathbb{C}^3$ in terms of a $SU(3)$-subgroup for $j \in \{1, 2, 3\}$ and an irreducible representation for $j \in \{2, 5, 7\}$. This difference may be understood from the fact that the first three generators

$$\lambda_1 = \begin{pmatrix} 0 & 1 & 0 \\ 1 & 0 & 0 \\ 0 & 0 & 0 \end{pmatrix} \lambda_2 = \begin{pmatrix} 0 & -i & 0 \\ i & 0 & 0 \\ 0 & 0 & 0 \end{pmatrix} \lambda_3 = \begin{pmatrix} 1 & 0 & 0 \\ 0 & -1 & 0 \\ 0 & 0 & 0 \end{pmatrix} \quad (2.113)$$

admit in contrast to the later, with

$$\lambda_5 = \begin{pmatrix} 0 & 0 & -i \\ 0 & 0 & 0 \\ i & 0 & 0 \end{pmatrix} \lambda_7 = \begin{pmatrix} 0 & 0 & 0 \\ 0 & 0 & -i \\ 0 & i & 0 \end{pmatrix} \quad (2.114)$$

an invariant decomposition into 2+1-dimensional subspaces in $\mathbb{C}^3$. We have therefore

$$R(i\sigma^j) = \lambda_j \begin{cases} j \in \{1,2,3\}, \text{ a reducible representation} \\ j \in \{2,5,7\}, \text{ an irreducible representation.} \end{cases} \quad (2.115)$$

By setting as fiducial vector

$$|0\rangle \equiv \begin{pmatrix} r_1 e^{i\beta_1} \\ r_2 e^{i\beta_2} \\ r_3 e^{i\beta_3} \end{pmatrix} \in \mathbb{C}_0^3, \quad r_j \in \mathbb{R}_0^+, \beta_j \in [0, 2\pi) \quad (2.116)$$

we may compute the pull-back tensors on the 3-dimensional manifold $SU(2) \cong S^3$ according to Theorem 2.4. We will distinguish in the following the dependence of the representation with a corresponding upper script notation $T^{1,2,3}$ for the reducible case, resp. $T^{2,5,7}$ for the irreducible case.

We may start by focusing on the linear part of the symmetric pull-back tensor

$$(L_{jk}^{1,2,3}) = \begin{pmatrix} \frac{r_1^2+r_2^2}{r_1^2+r_2^2+r_3^2} & 0 & 0 \\ 0 & \frac{r_1^2+r_2^2}{r_1^2+r_2^2+r_3^2} & 0 \\ 0 & 0 & \frac{r_1^2+r_2^2}{r_1^2+r_2^2+r_3^2} \end{pmatrix} \quad (2.117)$$

$$(L_{jk}^{2,5,7}) = \begin{pmatrix} \frac{r_1^2+r_2^2}{r_1^2-r_2^2+r_3^2} & \frac{\cos(\beta_{23})r_{23}}{r_1^2+r_2^2+r_3^2} & -\frac{\cos(\beta_{13})r_{13}}{r_1^2+r_2^2+r_3^2} \\ \frac{\cos(\beta_{23})r_{23}}{r_1^2+r_2^2+r_3^2} & 1-\frac{r_2^2}{r_1^2+r_2^2+r_3^2} & \frac{\cos(\beta_{12})r_{12}}{r_1^2+r_2^2+r_3^2} \\ -\frac{\cos(\beta_{13})r_{13}}{r_1^2+r_2^2+r_3^2} & \frac{\cos(\beta_{12})r_{12}}{r_1^2+r_2^2+r_3^2} & 1-\frac{r_1^2}{r_1^2+r_2^2+r_3^2} \end{pmatrix}, \quad (2.118)$$

where we observe that only the irreducible representation-dependent pull-back contains the phase differences associated to the 'interference terms'. When we focus on the other hand on the symmetric tensors $G^{1,2,3}$ and $G^{2,5,7}$ on the following pages, the information on the superposition of the state appears in both cases. The irreducible counterpart $G^{2,5,7}$ contains nevertheless again more information, in particular in terms of the individual difference on phase

differences
$$\beta_{abcb} := \beta_{ab} - \beta_{cb} \tag{2.119}$$

with $\beta_{ab} := \beta_a - \beta_b$ not contained in the 'ordinary' interference terms appearing in the pure density matrix ρ_0, as indicated in the previous section according to

$$\rho_0 := \frac{|0\rangle \langle 0|}{\langle 0 |0\rangle} = \frac{1}{r_1^2 + r_2^2 + r_3^2} \begin{pmatrix} r_1^2 & e^{i\beta_2 - i\beta_1} r_1 r_2 & e^{i\beta_3 - i\beta_1} r_1 r_3 \\ e^{i\beta_1 - i\beta_2} r_1 r_2 & r_2^2 & e^{i\beta_3 - i\beta_2} r_2 r_3 \\ e^{i\beta_1 - i\beta_3} r_1 r_3 & e^{i\beta_2 - i\beta_3} r_2 r_3 & r_3^2 \end{pmatrix}.$$

$$G^{1,2,3} = \begin{pmatrix} \dfrac{r_1^4 + r_2^4 + (r_1^2+r_2^2)r_3^2 - 2\cos(2\beta_{12})r_{12}^2}{(r_1^2+r_2^2+r_3^2)^2} & -\dfrac{2\sin(2\beta_{12})r_{12}^2}{(r_1^2+r_2^2+r_3^2)^2} & \dfrac{2\cos(\beta_{12})(r_2^2-r_1^2)r_{12}^2}{(r_1^2+r_2^2+r_3^2)^2} \\[2ex] \dfrac{2\cos(\beta_{12})(r_2^2-r_1^2)r_{12}}{(r_1^2+r_2^2+r_3^2)^2} & \dfrac{r_1^4+r_2^4+(r_1^2+r_2^2)r_3^2+2\cos(2\beta_{12})r_{12}^2}{(r_1^2+r_2^2+r_3^2)^2} & \dfrac{2\sin(\beta_{12})(r_2^2-r_1^2)r_{12}^2}{(r_1^2+r_2^2+r_3^2)^2} \\[2ex] \dfrac{2\sin(\beta_{12})r_{12}^2}{(r_1^2+r_2^2+r_3^2)^2} & \dfrac{2\sin(\beta_{12})(r_2^2-r_1^2)r_{12}^2}{(r_1^2+r_2^2+r_3^2)^2} & \dfrac{(r_1^2+r_2^2)r_3^2+4r_{12}^2}{(r_1^2+r_2^2+r_3^2)^2} \end{pmatrix} \quad (2.120)$$

$$G^{1,2,5,7} = \begin{pmatrix} \dfrac{r_1^4+r_2^4+(r_1^2+r_2^2)r_3^2+2\cos(2\beta_{12})r_{12}^2}{(r_1^2+r_2^2+r_3^2)^2} & \dfrac{(2\cos(\beta_{1312})-\cos(\beta_{23}))r_1^2+\cos(\beta_{23})(r_2^2+r_3^2))r_{23}}{(r_1^2+r_2^2+r_3^2)^2} & -\dfrac{(\cos(\beta_{13})r_1^2-(\cos(\beta_{13})-2\cos(\beta_{1232})r_2^2+\cos(\beta_{13}))r_3^2)r_{13}}{(r_1^2+r_2^2+r_3^2)^2} \\[2ex] \dfrac{(2\cos(\beta_{1312})-\cos(\beta_{23}))r_1^2+\cos(\beta_{23})(r_2^2+r_3^2))r_{23}}{(r_1^2+r_2^2+r_3^2)^2} & \dfrac{r_1^4+(r_2^2+2\cos(2\beta_{13})r_3^2)r_1^2+r_2^2(r_2^2+r_3^2)}{(r_1^2+r_2^2+r_3^2)^2} & \dfrac{\cos(\beta_{12})(r_1^2+r_2^2)-\cos(\beta_{12})-2\cos(\beta_{13232})r_3^2)r_{12}}{(r_1^2+r_2^2+r_3^2)^2} \\[2ex] -\dfrac{(\cos(\beta_{13})r_1^2-(\cos(\beta_{13})-2\cos(\beta_{1232}))r_2^2+\cos(\beta_{13})r_3^2)r_{13}}{(r_1^2+r_2^2+r_3^2)^2} & \dfrac{(\cos(\beta_{1,r})(r_1^2+r_2^2)-2\cos(\beta_{1323})r_3^2)r_{12}}{(r_1^2+r_2^2+r_3^2)^2} & \dfrac{r_2^4+r_3^4+2\cos(2\beta_{23})r_{23}^2+r_1^2(r_2^2+r_3^2)}{(r_1^2+r_2^2+r_3^2)^2} \end{pmatrix} \quad (2.121)$$

By focusing finally on the anti-symmetric part of the pull-back tensor we find:

$$(\Omega_{jk}^{1,2,3}) = \begin{pmatrix} 0 & \frac{r_2^2 - r_1^2}{r_1^2 + r_2^2 + r_3^2} & -\frac{2\sin(\beta_{12})r_{12}}{r_1^2 + r_2^2 + r_3^2} \\ \frac{r_2^2 - r_1^2}{r_1^2 + r_2^2 + r_3^2} & 0 & \frac{2\cos(\beta_{12})r_{12}}{r_1^2 + r_2^2 + r_3^2} \\ \frac{2\sin(\beta_{12})r_{12}}{r_1^2 + r_2^2 + r_3^2} & -\frac{2\cos(\beta_{12})r_{12}}{r_1^2 + r_2^2 + r_3^2} & 0 \end{pmatrix} \qquad (2.122)$$

$$(\Omega_{jk}^{2,5,7}) = \begin{pmatrix} 0 & \frac{\sin(\beta_{23})r_{23}}{r_1^2 + r_2^2 + r_3^2} & -\frac{\sin(\beta_{13})r_{13}}{r_1^2 + r_2^2 + r_3^2} \\ -\frac{\sin(\beta_{23})r_{23}}{r_1^2 + r_2^2 + r_3^2} & 0 & \frac{\sin(\beta_{12})r_{12}}{r_1^2 + r_2^2 + r_3^2} \\ \frac{\sin(\beta_{13})r_{13}}{r_1^2 + r_2^2 + r_3^2} & -\frac{\sin(\beta_{12})r_{12}}{r_1^2 + r_2^2 + r_3^2} & 0 \end{pmatrix}. \qquad (2.123)$$

Here it appears that it is the reducible rather than the irreducible representation dependent pulled back tensor, which contains an additional information on the state by means of the differences on the amplitudes

$$r_a^2 - r_b^2, \qquad (2.124)$$

being not contained in the state ρ_0.

Let us comment these results with a possible outlook on their applications. We have seen in the pull-back procedure of the previous section that a given defining representation will admit additional structures with increasing dimensions. Nevertheless one may consider lower dimensional group actions, whenever we know which particular information we shall extract from the quantum state associated to a higher dimensional Hilbert space.

2.3.2 Representations of U(1) on $\mathbb{C}^n$

In the most extreme case, we may consider the representations of an one dimensional Lie group, given by $U(1)$. In this case however, we will encounter only the 'diagonal elements' L_{jj}, G_{jj} and $\Omega_{jj} = 0$ of the higher dimensional representation-dependent pull-back tensor fields on 1-dimensional orbits. More specific, we will have the Lie algebra representations of the generator of $U(1)$ given by the imaginary unit i, where we may set without loss of generality

$$R^{(j)}(i) \equiv \sigma_j \in u^*(n) \qquad (2.125)$$

for each representation in a given Hilbert space $\mathcal{H} \cong \mathbb{C}^n$ and find

$$L = \left(\frac{\langle 0 | \sigma_j^2 | 0 \rangle}{\langle 0 | 0 \rangle} \right) \theta^j \odot \theta^j \qquad (2.126)$$

$$G = \left(\frac{\langle 0 | \sigma_j^2 | 0 \rangle}{\langle 0 | 0 \rangle} - \frac{\langle 0 | \sigma_j | 0 \rangle^2}{\langle 0 | 0 \rangle^2} \right) \theta^j \odot \theta^j. \qquad (2.127)$$

$$\Omega = 0 \qquad (2.128)$$

as pull-back tensor fields on $U(1) \cong S^1$.

2.3.3 Product representation of U(n)×U(n) on $\mathbb{C}^{n^2}$

A natural class of unitary subgroups arises by proceeding to composite quantum systems being realized by means of tensor products of Hilbert spaces, each of complex dimension n,

$$\mathcal{H} = \mathcal{H}_A \otimes \mathcal{H}_B \cong \mathbb{C}^n \otimes \mathbb{C}^n \cong \mathbb{C}^{n^2} \tag{2.129}$$

yielding Hilbert spaces of complex dimension n^2. Within the associated unitary group $U(n^2)$ acting on $\mathcal{H} \cong \mathbb{C}^{n^2}$, we encounter the unitary subgroup

$$U(n) \times U(n) \subset U(n^2). \tag{2.130}$$

To identify a well-defined action of

$$\mathcal{G} \equiv U(n) \times U(n) \tag{2.131}$$

on the tensor-product Hilbert space we may consider a product representation

$$\mathcal{G} = U(n) \times U(n) \to U(n^2) \tag{2.132}$$
$$g \equiv (g_A, g_B) \mapsto U(g) \equiv g_A \otimes g_B \tag{2.133}$$

and find:

Proposition 2.12. *Let* $\mathcal{H} := \mathcal{H}_A \otimes \mathcal{H}_B \cong \mathbb{C}^n \otimes \mathbb{C}^n$ *be a composite Hilbert space, let* $\{i\sigma_j\}_{0 \le j \le n^2-1}$ *be a trace-orthonormal basis of the Lie-algebra* $u(n)$ *and let* U *be a product representation of* $\mathcal{G} = U(n) \times U(n)$ *on the Hilbert space* $(\mathcal{H}, \tau_{\mathcal{H}_0})$. *For a given* $U(n) \times U(n) \cdot_U |0\rangle$-*orbit* $\mathcal{O}$ *there exists on* $U(n) \times U(n)$ *a pull-back tensor field* $\tau_{U(n) \times U(n)}$, *having a coefficient matrix*

$$(T_{jk}) = \begin{pmatrix} (G^A_{jk}) & (G^{AB}_{jk}) \\ (G^{AB}_{jk}) & (G^B_{jk}) \end{pmatrix} + i \begin{pmatrix} (\Omega^A_{[jk]}) & 0 \\ 0 & (\Omega^B_{[jk]}) \end{pmatrix}, \tag{2.134}$$

with

$$G^A_{(jk)} = \frac{\langle 0|\,[\sigma_j,\sigma_k]_+ \otimes \mathbf{1}\,|0\rangle}{\langle 0|0\rangle} - \frac{\langle 0|\,\sigma_j \otimes \mathbf{1}\,|0\rangle\,\langle 0|\,\sigma_k \otimes \mathbf{1}\,|0\rangle}{\langle 0|0\rangle^2}$$

$$\Omega^A_{[jk]} = \frac{\langle 0|\,[\sigma_j,\sigma_k]_- \otimes \mathbf{1}\,|0\rangle}{\langle 0|0\rangle}, \quad \text{for } 1 \le j,k \le n^2; \tag{2.135}$$

$$G^B_{(jk)} = \frac{\langle 0|\,\mathbf{1} \otimes [\sigma_{j-n^2},\sigma_{k-n^2}]_+\,|0\rangle}{\langle 0|0\rangle} - \frac{\langle 0|\,\mathbf{1} \otimes \sigma_{j-n^2}\,|0\rangle\,\langle 0|\,\mathbf{1} \otimes \sigma_{k-n^2}\,|0\rangle}{\langle 0|0\rangle^2}$$

$$\Omega^B_{[jk]} = \frac{\langle 0|\,\mathbf{1} \otimes [\sigma_{j-n^2},\sigma_{k-n^2}]_-\,|0\rangle}{\langle 0|0\rangle}, \quad \text{for } n^2+1 \le j,k \le 2n^2; \text{ and} \tag{2.136}$$

$$G^{AB}_{jk} = \frac{\langle 0|\,\sigma_j \otimes \sigma_{k-n^2}\,|0\rangle}{\langle 0|0\rangle} - \frac{\langle 0|\,\sigma_j \otimes \mathbf{1}\,|0\rangle\,\langle 0|\,\mathbf{1} \otimes \sigma_{k-n^2}\,|0\rangle}{\langle 0|0\rangle^2} \tag{2.137}$$

for $1 \le j \le n^2, n^2+1 \le k \le 2n^2$.

Proof. The representation (2.133) admits a factorization

$$U(g) = g_A \otimes g_B \tag{2.138}$$
$$= (g_A \otimes \mathbf{1})(\mathbf{1} \otimes g_B) \tag{2.139}$$
$$:= U_A(g_A) U_B(g_B) \tag{2.140}$$

into two single defining representations of $g_s \in U(n)$ on $\mathcal{H}_s \cong \mathbb{C}^n$ being tensored with the identity according to

$$U_s(g_s) := \begin{cases} g_A \otimes \mathbf{1} & \text{for } s = A \\ \mathbf{1} \otimes g_B & \text{for } s = B. \end{cases} \tag{2.141}$$

In this way we encounter a commuting action

$$[U_A(g_A), U_B(g_B)] = 0, \tag{2.142}$$

since we have

$$(g_A \otimes \mathbf{1})(\mathbf{1} \otimes g_B) = (\mathbf{1} \otimes g_B)(g_A \otimes \mathbf{1}). \tag{2.143}$$

With $\varphi^j \in [0, 2\pi)$ and $\{X_j\}_{j \le 2n^2}$, a Lie algebra basis of $u(n) \oplus u(n)$, we conclude within the Lie algebra representation expansion of the Lie group representation

$$\begin{array}{ccc} U(n) \times U(n) & \xrightarrow{U} & U(n^2) \\ \exp \uparrow & & \exp \uparrow \\ u(n) \oplus u(n) & \xrightarrow{iR} & u(n^2), \end{array}$$

that the corresponding infinitesimal action of $U(g) = U(g_A)U(g_B)$ resp.

$$U(g) = \prod_{j=0}^{2n^2} e^{i\varphi^j R(X_j)} = \prod_{j=1}^{n^2} e^{i\varphi^j R_A(X_j)} \prod_{j=n^2+1}^{2n^2} e^{i\varphi^j R_B(X_j)} \tag{2.144}$$

decomposes into a sum of the infinitesimal action of the defining representation tensored by the identity according to

$$\sum_{j=1}^{2n^2} \varphi^j R(X_j) = \sum_{j=1}^{n^2} \varphi^j R_A(i\sigma_j) + \sum_{j=n^2+1}^{2n^2} \varphi^j R_B(i\sigma_j). \tag{2.145}$$

$$:= \sum_{j=1}^{n^2} \varphi^j \sigma_j \otimes \mathbb{1} + \sum_{j=n^2+1}^{2n^2} \varphi^j \mathbb{1} \otimes \sigma_{j-n^2}. \tag{2.146}$$

Hence, by means of the basis of n^2 trace-orthonormal Hermitian matrices $\sigma_j \in u^*(n)$, with $1 \leq j \leq n^2 - 1$ and $\sigma_0 = \mathbb{1}$, which we have used in Proposition 2.10 in the previous section 2.2 within the defining representation on a n-dimensional Hilbert space, we identify the Hermitian matrices

$$R(X_j) = \begin{cases} R_A(X_j) = \sigma_j \otimes \mathbb{1} & \text{for } 1 \leq j \leq n^2 \\ R_B(X_j) = \mathbb{1} \otimes \sigma_{j-n^2} & \text{for } n^2 + 1 \leq j \leq 2n^2, \end{cases} \tag{2.147}$$

defining the Lie algebra representation of the basis $\{X_j\}_{j \leq 2n^2}$ of the Lie algebra $u(n) \oplus u(n)$. These Hermitian matrices imply three different classes of symmetric and anti-symmetric combinations:

$$[R(X_j), R(X_k)]_+ =$$

$$\begin{cases} [(\sigma_j \otimes \mathbb{1}), (\sigma_k \otimes \mathbb{1})]_+ = [\sigma_j, \sigma_k]_+ \otimes \mathbb{1} & \text{for } 1 \leq j, k \leq n^2 \\ [(\mathbb{1} \otimes \sigma_{j-n^2}), (\mathbb{1} \otimes \sigma_{k-n^2})]_+ = \mathbb{1} \otimes [\sigma_{j-n^2}, \sigma_{k-n^2}]_+ & \text{for } n^2 + 1 \leq j, k \leq 2n^2 \\ [(\sigma_j \otimes \mathbb{1}), (\mathbb{1} \otimes \sigma_{k-n^2})]_+ = \sigma_j \otimes \sigma_{k-n^2} & \text{otherwise.} \end{cases} \tag{2.148}$$

$$[R(X_j), R(X_k)]_- =$$

$$\begin{cases} [(\sigma_j \otimes \mathbb{1}), (\sigma_k \otimes \mathbb{1})]_- = [\sigma_j, \sigma_k]_- \otimes \mathbb{1} & \text{for } 1 \leq j, k \leq n^2 \\ [(\mathbb{1} \otimes \sigma_{j-n^2}), (\mathbb{1} \otimes \sigma_{k-n^2})]_- = \mathbb{1} \otimes [\sigma_{j-n^2}, \sigma_{k-n^2}]_- & \text{for } n^2 + 1 \leq j, k \leq 2n^2 \\ [(\sigma_j \otimes \mathbb{1}), (\mathbb{1} \otimes \sigma_{k-n^2})]_- = 0 & \text{otherwise.} \end{cases} \tag{2.149}$$

With Theorem 2.4 and a decomposition of the coefficient matrix $(T_{(jk)})$ within the pull-back tensor $\tau_{U(n) \times U(n)}$ into sub-matrices

$$(T_{(jk)}) := \begin{pmatrix} (T^A_{(jk)}) & (T^{AB}_{(jk)}) \\ (T^{AB}_{(jk)}) & (T^B_{(jk)}) \end{pmatrix} \tag{2.150}$$

according to

$$(T^A_{(jk)}) \subset (T_{(jk)}) :\Leftrightarrow \quad 1 \leq j, k \leq n^2 \qquad (2.151)$$
$$(T^B_{(jk)}) \subset (T_{(jk)}) :\Leftrightarrow \quad n^2 + 1 \leq j, k \leq 2n^2 \qquad (2.152)$$
$$(T^{AB}_{jk}) \subset (T_{(jk)}) :\Leftrightarrow \quad \text{otherwise,} \qquad (2.153)$$

we end up with the statement. $\square$

3 Application on the separability problem in composite systems $\mathcal{H}_A \otimes \mathcal{H}_B$

In the following we will show a possible application of the pull-back structures of the previous chapter in a concrete problem being of particular interest in quantum information and quantum computation. It is given by the so-called *separability problem* for pure quantum states in a composite bi-partite system. Consider for this purpose the following definition:

Definition 3.1 (Separable and entangled vectors). *A vector* $|\psi\rangle \in \mathcal{H}_A \otimes \mathcal{H}_B$ *is called separable if*

$$|\psi\rangle = |\psi_A\rangle \otimes |\psi_B\rangle \tag{3.1}$$

for $|\psi_A\rangle \in \mathcal{H}_A$ *and* $|\psi_B\rangle \in \mathcal{H}_B$, *otherwise entangled.*

The notion of entanglement is the most characteristic feature, which distinguishes quantum from classical physics[5]. On the kinematical level it is implemented by means of the tensor product of Hilbert spaces, which implies a different increase of dimensions in respect to the cartesian product, commonly used in classical mechanics to formulate composite systems, where one finds

$$\text{Dim}_{\mathbb{C}}(\mathbb{C}^n \otimes \mathbb{C}^n) = n^2 \tag{3.2}$$

in contrast to

$$\text{Dim}_{\mathbb{C}}(\mathbb{C}^n \times \mathbb{C}^n) = 2n. \tag{3.3}$$

It is this distinguished scaling of dimensions, which usually motivates the idea on quantum information and quantum computation, exceeding the possibilities of classical computation [74]. Crucially, also on the projective Hilbert space of pure quantum states, where we will not have a linear structure available to define a tensor product, we encounter a distinguished topology from a cartesian product of two single projective Hilbert spaces unless we are considering separable vectors on the associated Hilbert space. The separability problem concerns on the question whether a given state 'lives' in such a cartesian product space of two single projective Hilbert spaces or not. The traditional approach to this problem is of operational nature and works as follows.

3.1 Traditional algebraic approach

For a given finite dimensional product Hilbert space

$$\mathcal{H}_A \otimes \mathcal{H}_B \cong \mathbb{C}^n \otimes \mathbb{C}^n \cong \mathbb{C}^{n^2} \tag{3.4}$$

---

[5]The notion of interference is also often cited in the literature separated from entanglement as a 'typical quantum phenomena'. However as we remarked before, the latter phenomena appears indeed also in classical theories, like in electromagnetism. On the other hand it appears that interference of states associated to tensor product Hilbert spaces is closely related to entanglement. For an intrinsic approach on both phenomena we refer to [58].

one may start by introducing an arbitrary orthogonal tensor product basis $\{|e_j\rangle \otimes |e_k\rangle\}_{j,k \in J}$ on $\mathcal{H}_A \otimes \mathcal{H}_B$. In this way any given vector $|\psi\rangle \in \mathcal{H}_A \otimes \mathcal{H}_B$ is expressed in a corresponding expansion

$$|\psi\rangle = \sum_{j=1}^{n} \sum_{k=1}^{n} c^{jk} |e_j\rangle \otimes |e_k\rangle. \qquad (3.5)$$

The complex coefficients c^{jk} of the vector can be reorganized in this regard as coefficients of a complex $n \times n$-matrix

$$(c^{jk}) := C_\psi \in \mathbb{C}^{n \times n}. \qquad (3.6)$$

Here we observe that any complex matrix admits a singular value decomposition

$$C_\psi \mapsto UC_\psi V^\dagger = \text{Diag}(\sqrt{\lambda_1}, \sqrt{\lambda_2}, ..) := \Lambda, \qquad (3.7)$$

$$\text{with } (U, V) \in U(n) \times U(n), \qquad (3.8)$$

giving rise to a diagonal matrix Λ with positive and real valued entries $\lambda_j \geqslant 0$ related to the eigenvalues of the positive Hermitian matrix $C_\psi^\dagger C_\psi$. This transformation induces a new basis $\{|e'_j\rangle \otimes |d'_j\rangle\}_{j \in J'}$ related to the old basis $\{|e_j\rangle \otimes |e_k\rangle\}_{j,k \in J}$ by a unitary transformation yielding a corresponding new expansion of the same state according to

$$|\psi\rangle = \sum_{j}^{n} \sqrt{\lambda_j} |e'_j\rangle \otimes |d'_j\rangle. \qquad (3.9)$$

This particular expansion is the so-called *Schmidt decomposition* of a given state (see e.g. [2]), where we observe that one has only one sum to take into account – in contrast to the case of an arbitrary tensor product basis expansion (3.5). Moreover, we see that the number of complex coefficients reduces to a lower number of coefficients, called the *Schmidt coefficients* λ_j, admitting – as being related to eigenvalues of $C_\psi^\dagger C_\psi$ – the property[6]

$$\lambda_j \in \mathbb{R}^+. \qquad (3.10)$$

By organizing the Schmidt-coefficients as a tuple of real numbers

$$\vec{\lambda} := (\lambda_1, \lambda_2, ..) \in (\mathbb{R}^+)^n \subset \mathbb{R}^n \qquad (3.11)$$

embedded in a real n-dimensional affine space, one finds in the case of normalized states

$$\langle \psi | \psi \rangle = 1 \Rightarrow \sum_{j}^{n} \lambda_j = 1, \qquad (3.12)$$

implying that the Schmidt-coefficients provide a parametrization of points

$$\vec{\lambda} \in \Delta_{n-1} \subset (\mathbb{R}^+)^n \qquad (3.13)$$

---

[6] We use the convention $\mathbb{R}_0^+ := \mathbb{R}^+ - \{0\}$, i.e. $\mathbb{R}^+$ denotes here the real positive numbers including zero.

on a polytope with n vertices, resp. a n-1-simplex $\triangle_{n-1}$. In this way one arrives to the conclusion:

Proposition 3.2. *For a given vector* $|\psi\rangle \in \mathcal{H}_A \otimes \mathcal{H}_A \cong \mathbb{C}^n \otimes \mathbb{C}^n$ *there exists a basis, such that*

$$|\psi\rangle = \sum_{j=1}^{n} \sqrt{\lambda_j} |e'_j\rangle \otimes |d'_j\rangle \tag{3.14}$$

holds with $\lambda_j \in \mathbb{R}^+$.

In other words: Once a Schmidt-decomposition (3.14) of a given vector is found, the solution of the separability problem is reached:

Corollary 3.3. *A normalized vector* $|\psi\rangle \in \mathcal{H}_A \otimes \mathcal{H}_B \cong \mathbb{C}^n \otimes \mathbb{C}^n$ *is separable*

$$\Leftrightarrow \quad \text{There is only one non-vanishing } \lambda_j = 1, \tag{3.15}$$
$$\Leftrightarrow \quad \vec{\lambda} \text{ is a vertex of a } n\text{-1-simplex } \triangle_{n-1}. \tag{3.16}$$

However, the associated singular value decomposition (3.7) turns out not to be an easy computational task, in particular when one encounters increasing dimensions. We may nevertheless encounter criteria with lower computational effort by relating vectors in the composite Hilbert space to their associated *reduced density states*, being defined by the partial traces of the pure state $\rho_\psi := |\psi\rangle\langle\psi|$ by

$$\rho_\psi^A := \text{Tr}_B(|\psi\rangle\langle\psi|), \tag{3.17}$$

$$\rho_\psi^B := \text{Tr}_A(|\psi\rangle\langle\psi|). \tag{3.18}$$

We recall the notion of a partial trace in a basis independent formulation: We say ρ_ψ^A and ρ_ψ^B are reduced density states of the pure state ρ_ψ iff

$$\text{Tr}(\rho_\psi\, X \otimes \mathbb{1}) = \text{Tr}(\rho_\psi^A\, X) \tag{3.19}$$

$$\text{Tr}(\rho_\psi\, \mathbb{1} \otimes X) = \text{Tr}(\rho_\psi^B\, X), \tag{3.20}$$

holds for any Hermitian matrix $X \in \mathfrak{u}^*(\mathcal{H}_s)$. Here one finds

Proposition 3.4. *For a given normalized vector in* $|\psi\rangle \in \mathcal{H}_A \otimes \mathcal{H}_B \cong \mathbb{C}^n \otimes \mathbb{C}^n$, *following statements are equivalent:*

$$|\psi\rangle \text{ is separable} \tag{a}$$

$$\text{Tr}((\rho_\psi^A)^2) = 1 \tag{b}$$

$$(\rho_\psi^A)^2 = \rho_\psi^A \tag{c}$$

$$\rho_\psi = \rho_\psi^A \otimes \rho_\psi^B \tag{d}$$

$$\text{Tr}(\rho_\psi \sigma_j \otimes \sigma_k) = \text{Tr}(\rho_\psi^A \sigma_j)\text{Tr}(\rho_\psi^B \sigma_k), \tag{e}$$

for all $\sigma_j \in \{\sigma_j\}_{0 \leq j \leq n^2-1}$, denoting a trace-orthonormal basis of the space of Hermitian matrices $u^*(n)$.

Proof. We show first the equivalence between (a), (b), (c):

(b) $\Leftrightarrow$ (a): Given a vector $|\psi\rangle \in \mathcal{H}_A \otimes \mathcal{H}_B \cong \mathbb{C}^n \otimes \mathbb{C}^n$, there will exists a Schmidt-decomposition

$$|\psi\rangle = \sum_j^n \sqrt{\lambda_j} |e'_j\rangle \otimes |d'_j\rangle \tag{3.21}$$

into Schmidt-coefficients $\lambda_j \in \mathbb{R}^+$, which are unique up to ordering, where $\{|e'_j\rangle \otimes |d'_j\rangle\}_{j \in J}$ denote an orthonormal basis on $\mathcal{H}_A \otimes \mathcal{H}_B$. The partial trace of the associated pure density state

$$\rho_\psi = |\psi\rangle\langle\psi| = \sum_{j,k} \sqrt{\lambda_j}\sqrt{\lambda_k} |e'_j\rangle\langle e'_k| \otimes |d'_j\rangle\langle d'_k| \tag{3.22}$$

reads

$$\rho_\psi^A = \mathrm{Tr}_A(|\psi\rangle\langle\psi|) = \sum_{j,k} \sqrt{\lambda_j}\sqrt{\lambda_k} \langle e'_l|e'_j\rangle\langle e'_k|e'_l\rangle \otimes |d'_j\rangle\langle d'_k|$$

$$= \sum_{j,k} \sqrt{\lambda_j}\sqrt{\lambda_k} \delta_{jl}\delta_{kl} |d'_j\rangle\langle d'_k|$$

$$= \sum_{j,k} \sqrt{\lambda_j}\sqrt{\lambda_k} \delta_{jk} |d'_j\rangle\langle d'_k|$$

$$= \sum_j \lambda_j |d'_j\rangle\langle d'_j|. \tag{3.23}$$

Here one finds

$$\mathrm{Tr}((\rho_\psi^A)^2) = \mathrm{Tr}(\sum_j \lambda_j^2 |d'_j\rangle\langle d'_j|) = \sum_j \lambda_j^2. \tag{3.24}$$

Hence, (b) holds iff there exists only one non-vanishing Schmidt-coefficient $\lambda_j = 1$. But according to (3.15), this is equivalent with (a).

(a) $\Rightarrow$ (c): Follows directly by setting only one non-vanishing Schmidt-coefficient $\lambda_j = 1$ in (3.23).

(c) $\Rightarrow$ (b): Follows directly from the normalization

$$\mathrm{Tr}(\rho_\psi^A) = 1. \tag{3.25}$$

Next we show the equivalence between (a), (d), (e):

(a) $\Leftrightarrow$ (d): According to (3.23) one finds

$$\rho_\psi^B = \sum_j \lambda_j |e'_j\rangle\langle e'_j|. \tag{3.26}$$

On the other hand we have

$$\rho_\psi = |\psi\rangle\langle\psi| = |e'_j\rangle\langle e'_j| \otimes |d'_j\rangle\langle d'_j| \tag{3.27}$$

iff the vector is separable.

(d) ⇒ (e): Follows directly from

$$\text{Tr}((\rho_\psi^A \otimes \rho_\psi^B)\sigma_j \otimes \sigma_k) = \text{Tr}(\rho_\psi^A \sigma_j \otimes \rho_\psi^B \sigma_k). \tag{3.28}$$

(e) ⇒ (d): To show the inverse direction, we express any Hermitian matrix, including a bipartite pure density state $\rho_\psi \in \mathcal{U}^*(\mathbb{C}^n \otimes \mathbb{C}^n)$ and the reduced density states $\rho_\psi^A, \rho_\psi^B \in \mathcal{U}^*(\mathbb{C}^n)$, in terms of the Bloch-representation expansion (see e.g. [28]),

$$\rho_\psi^A = \frac{1}{n}(\sigma_0 + n_j \sigma_j) \tag{3.29}$$

$$\rho_\psi^B = \frac{1}{n}(\sigma_0 + m_k \sigma_k) \tag{3.30}$$

$$\rho_\psi \equiv \frac{1}{n^2}(\sigma_0 \otimes \sigma_0 + n_j \sigma_j \otimes \sigma_0 + m_k \sigma_0 \otimes \sigma_k + t_{jk} \sigma_j \otimes \sigma_k). \tag{3.31}$$

with

$$n_j := \text{Tr}(\rho_\psi^A \sigma_j) = \text{Tr}(\rho_\psi \sigma_j \otimes \mathbb{1}) \tag{3.32}$$

$$m_k := \text{Tr}(\rho_\psi^B \sigma_k) = \text{Tr}(\rho_\psi \mathbb{1} \otimes \sigma_k) \tag{3.33}$$

$$t_{jk} := \text{Tr}(\rho_\psi \sigma_j \otimes \sigma_k). \tag{3.34}$$

From (e) it follows

$$t_{jk} - n_j m_k = 0 \tag{3.35}$$

and therefore

$$\rho_\psi = \frac{1}{n^2}(\sigma_0 \otimes \sigma_0 + n_j \sigma_j \otimes \sigma_0 + m_k \sigma_0 \otimes \sigma_k + n_j m_k \sigma_j \otimes \sigma_k)$$

$$= (\frac{1}{n}(\sigma_0 + n_j \sigma_j)) \otimes (\frac{1}{n}(\sigma_0 + m_k \sigma_k)). \tag{3.36}$$

□

3.2 A geometric approach

Given a tuple $\vec{\lambda}$ of Schmidt coefficients λ_j we may perform the inverse transformation which has been used to provide a singular value decomposition of a state $|\psi\rangle$ according to

$$\vec{\lambda} :\Leftrightarrow \text{Diag}(\sqrt{\lambda_1}, \sqrt{\lambda_2}, ..) := \Lambda \mapsto U^\dagger \Lambda V = C_\psi \tag{3.37}$$

$$\text{with } (U, V) \in U(n) \times U(n), \tag{3.38}$$

to 'reconstruct' the initial complex coefficient matrix C_ψ in a given basis. Here it becomes clear that we may arrive to a whole family of states $|\psi'\rangle$, each with a complex coefficient matrix $C_{\psi'}$, parametrized by transformations $(U', V') \in U(n) \times U(n)$ for a given tuple $\vec{\lambda}$ of Schmidt coefficients encoded by Λ. This suggests to introduce the following notion [2]:

Definition 3.5 (Schmidt equivalence class). *Let C_ψ be the complex coefficient matrix associated to a vector $|\psi\rangle \in \mathcal{H}_A \otimes \mathcal{H}_B$ and $\Lambda = U C_\psi V^\dagger$ its decomposition in a tuple $\vec{\lambda}$ of Schmidt-coefficients induced by $(U, V) \in U(n) \times U(n)$. Any two vectors $|\psi\rangle, |\psi'\rangle \in \mathcal{H}_A \otimes \mathcal{H}_B$ are said to be in a Schmidt equivalence class $[\psi]_{\vec{\lambda}}$ whenever the following equivalence relation holds:*

$$|\psi\rangle \sim |\psi'\rangle :\Leftrightarrow$$

$$\exists (U', V') \in U(n) \times U(n) :$$

$$C_{\psi'} = U'^\dagger U C_\psi V^\dagger V' = U'^\dagger \Lambda V'. \tag{3.39}$$

Remark 3.6. *In the following we follow the convention, that the tuple $\vec{\lambda}$ is ordered by a sequence of decreasing Schmidt-coefficients. In this way we shall denote $\triangle$ from now on as the set of ordered Schmidt tuples, given by an identification on the simplex of unordered Schmidt tuples via unitary permutations.*

By setting

$$C_{\psi'} = U'^\dagger U C_\psi V^\dagger V' \equiv U_A C_\psi V_B \tag{3.40}$$

with

$$U_A \equiv U'^\dagger U \in U(n) \tag{3.41}$$

$$V_B \equiv V^\dagger V' \in U(n), \tag{3.42}$$

it follows that $|\psi\rangle$ and $|\psi'\rangle$ live on a common $U(n) \times U(n)$-orbit. In this way we may recover a Schmidt-equivalence class $[\psi]_{\vec{\lambda}}$ by applying the whole family of transformations given by $U(n) \times U(n)$ on a fiducial state $|0\rangle \in \mathcal{H}_A \otimes \mathcal{H}_B$ with a coefficient matrix C_0 and Schmidt-coefficients $\vec{\lambda}$. Here we will find a subset of states parametrized by points on

$$[\psi]_{\vec{\lambda}} \cong \frac{U(n) \times U(n)}{\mathcal{G}_{0,\vec{\lambda}}}, \tag{3.43}$$

a *homogeneous space* with isotropy group

$$\mathcal{G}_{\mathbb{C},\vec{\lambda}} := \{(U', V') \in U(n) \times U(n) |$$

$$U'\Lambda V'^\dagger = \Lambda = UC_0 V^\dagger, (U, V) \in U(n) \times U(n)\}. \tag{3.44}$$

This suggests to identify all states on the projective Hilbert space which have the same Schmidt coefficients

$$\vec{\lambda} \equiv \vec{\lambda}_\psi = \vec{\lambda}_{\psi'} \tag{3.45}$$

by means of an equivalence relation

$$|\psi'\rangle \sim |\psi\rangle :\Leftrightarrow |\psi'\rangle = e^{i\alpha} U_A \otimes V_B |\psi\rangle, \tag{3.46}$$

on the subset of normalized states in the Hilbert space, which yields a *Schmidt equivalence class* $[\psi]_{\vec{\lambda}}^{U(1)}$ defined in the projective Hilbert space.

These equivalence classes turn out to be related to the above homogeneous spaces (3.43), providing a stratification

$$\mathcal{P}(\mathbb{C}^n \otimes \mathbb{C}^n) \cong \bigcup_{\vec{\lambda} \in \Delta_{n-1}} [\psi]_{\vec{\lambda}}^{U(1)} \tag{3.47}$$

of the projective Hilbert space [51,75], where we have

$$[\psi]_{\vec{\lambda}}^{U(1)} \cong \frac{U(n) \times U(n)}{\mathcal{G}_{0,\vec{\lambda}}^{U(1)}} \tag{3.48}$$

with the 'extended' isotropy group

$$\mathcal{G}_{0,\vec{\lambda}}^{U(1)} := \{(U', V') \in U(n) \times U(n) |$$

$$U'\Lambda V'^\dagger = e^{i\alpha}\Lambda = UC_0 V^\dagger, (U, V) \in U(n) \times U(n)\}. \tag{3.49}$$

We may illustrate the Schmidt equivalence classes for $n = 2$. The space of normalized bi-partite 2-level states is defined in this case by the base manifold within the U(1)-fibration

$$\mathbb{C}^2 \otimes \mathbb{C}^2 \supset S^7 \xrightarrow{U(1)} \mathcal{P}(\mathbb{C}^2 \otimes \mathbb{C}^2) = \mathbb{C}P^3, \tag{3.50}$$

yielding the complex projective Hilbert space $\mathbb{C}P^3$. The later provides in real terms according to the Majorana stelar representation

$$\mathbb{C}P^3 \cong (S^2)^{\times 3}/S_3, \tag{3.51}$$

a configuration space of 3 unordered points on a 2-sphere [56,62] and admits a stratification

[51, 75]
$$\mathbb{C}P^3 \cong \bigcup_{\vec{\lambda} \in \Delta_1} [\psi]_{\vec{\lambda}}^{U(1)} \tag{3.52}$$

into Schmidt equivalence classes

$$[\psi]_{(1,0)}^{U(1)} \cong SU(2)/U(1) \times SU(2)/U(1), \tag{3.53}$$

$$[\psi]_{(a,b)}^{U(1)} \cong SU(2)/U(1) \times SU(2)/\mathbb{Z}_2, \tag{3.54}$$

$$[\psi]_{(\frac{1}{2},\frac{1}{2})}^{U(1)} \cong SU(2)/\mathbb{Z}_2, \tag{3.55}$$

which may be illustrated in terms of real manifolds as

$$[\psi]_{(1,0)}^{U(1)} \cong S^2 \times S^2, \tag{3.56}$$

$$[\psi]_{(a,b)}^{U(1)} \cong S^2 \times (S^3/\mathbb{Z}_2), \tag{3.57}$$

$$[\psi]_{(\frac{1}{2},\frac{1}{2})}^{U(1)} \cong S^3/\mathbb{Z}_2. \tag{3.58}$$

The subset of separable states given by the equivalence class with $\vec{\lambda} = (1,0)$ provides a Cartesian product of two projective Hilbert spaces, resp. Bloch spheres S^2 associated to a single 'qubit' system $\mathcal{H}_A \cong \mathcal{H}_B \cong \mathbb{C}^2$. On the other hand we may observe that the last class admitting equal distributed Schmidt coefficients $(\frac{1}{2}, \frac{1}{2})$ is distinguished from both, the separable states and the generic classes of entangled states with arbitrary distributed Schmidt coefficients (a, b). It is a 3-dimensional submanifold of the 6-dimensional space of states $\mathbb{C}P^3$, which is Kählerian and therefore also symplectic. Hence, the Schmidt-equivalence class $S^3/\mathbb{Z}_2$ provides an instance of a *Lagrangian submanifold* [8], i.e. a submanifold of half dimension of a symplectic manifold where the restriction of the symplectic form vanishes. This suggests to introduce the following definition:

Definition 3.7 (Lagrangian entanglement). *Let $\mathcal{O}$ denote a Schmidt equivalence class of entangled states associated to a Hilbert space $\mathcal{H}_A \otimes \mathcal{H}_B$. A state $|\psi\rangle \in \mathcal{O}$ is called Lagrangian entangled if $\mathcal{O}$ is a Lagrangian submanifold in $\mathcal{H}_A \otimes \mathcal{H}_B$.*

We may ask whether the topological properties of a Schmidt-equivalence class provide us also in arbitrary dimension some information whether the state is separable, entangled or Lagrangian entangled. To get an idea on the stratification of the space of states in Schmidt-equivalence classes, we may consider a picture as indicated here by figure 1.

Every Schmidt-equivalence $[\psi]_{\vec{\lambda}}$ class provides a homogeneous space living over a point of the simplex Δ parametrized by ordered Schmidt-coefficient tuples $\vec{\lambda}$. The arrow within one of the equivalence classes $[\psi]_{\vec{\lambda}}$ over the tuple $\vec{\lambda}_\psi \in \Delta$ denotes here a transformation induced by the unitary subgroup $U(n) \times U(n)$ from one state ψ to another ψ'. By identifying ψ as the representative element of the equivalence class in a Schmidt-decomposition basis, it may become

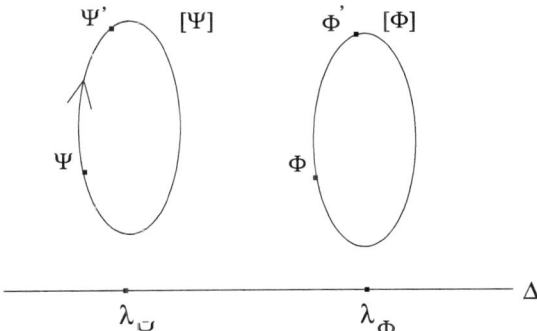

Figure 1: Global Schmidt decomposition map...

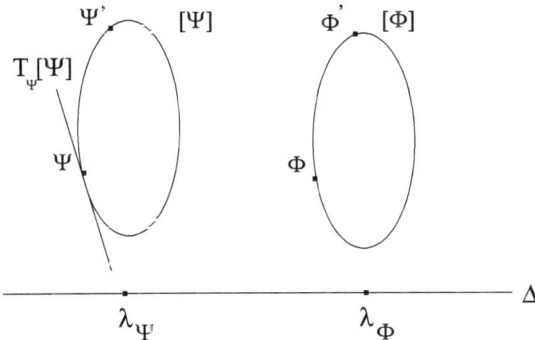

Figure 2: ...vs. local analysis of a Schmidt equivalence class

clear that any transformation to a generic state ψ' within the same equivalence class takes into account the singular value decomposition of the latter. Since the property of separability is a global property, we may replace such a *global* Schmidt-decomposition-map by a *local* analysis of the Schmidt-equivalence class in particular by focusing only on the tangent space

$$T_\psi[\psi]_{\vec{\lambda}} \cong T_{\psi'}[\psi]_{\vec{\lambda}}$$

over the corresponding given state as indicated in figure 2. By observing that each Schmidt-equivalence class provides an instance of an $U(n) \times U(n)$-generated orbit of quantum states being embedded in a tensor product Hilbert space $\mathcal{H}_A \otimes \mathcal{H}_B$, we may apply our previous pull-back tensor construction for this purpose and find:

Theorem 3.8. *Let $|0\rangle \in \mathcal{H}_A \otimes \mathcal{H}_B$ be a normalized vector. For a given $U(n) \times U(n) \cdot_U |0\rangle$-orbit embedded by a product representation in a Hilbert space $(\mathcal{H}_A \otimes \mathcal{H}_B, \tau_{\mathcal{H}_0})$ there exists an induced pull-back tensor field $\tau_{U(n) \times U(n)}$ on $U(n) \times U(n)$ which admits*

(a) a direct sum decomposition $\tau_{U(n)} \oplus \tau_{U(n)}$ iff $|0\rangle$ is separable,

(b) a maximal degenerate antisymmetric part iff $|0\rangle$ belongs to a Schmidt-equivalence class with equally distributed Schmidt-coefficients;

Proof. (a) According Proposition 2.12 one finds a pull-back tensor $\tau_{U(n) \times U(n)}$, having a coefficient matrix

$$(T_{jk}) = \begin{pmatrix} (G^A_{(jk)}) & (G^{AB}_{(jk)}) \\ (G^{AB}_{(jk)}) & (G^B_{(jk)}) \end{pmatrix} + i \begin{pmatrix} (\Omega^A_{[jk]}) & 0 \\ 0 & (\Omega^B_{[jk]}) \end{pmatrix}, \quad (3.59)$$

where

$$G^{AB}_{(jk)} = \frac{\langle 0|\, \sigma_j \otimes \sigma_{k-n^2}\, |0\rangle}{\langle 0|0\rangle} - \frac{\langle 0|\, \sigma_j \otimes \mathbb{1}\, |0\rangle \langle 0|\, \mathbb{1} \otimes \sigma_{k-n^2}\, |0\rangle}{\langle 0|0\rangle^2}. \quad (3.60)$$

Hence, the pull-back tensor splits into the direct sum of two tensor fields each defined on $U(n)$ iff $G^{AB}_{(jk)} = 0$. With part (e) in Proposition 3.4 one concludes then the statement (a).

(b) For the anti-symmetric coefficients we find according Proposition 2.12

$$\Omega^A_{[jk]} = \frac{\langle 0|\, [\sigma_j, \sigma_k]_- \otimes \mathbb{1}\, |0\rangle}{\langle 0|0\rangle}, \quad (3.61)$$

$$\Omega^B_{[jk]} = \frac{\langle 0|\, \mathbb{1} \otimes [\sigma_{j-n^2}, \sigma_{k-n^2}]_-\, |0\rangle}{\langle 0|0\rangle}, \quad (3.62)$$

implying in both cases

$$\Omega^s_{[jk]} = \text{Tr}(\text{Tr}_s(\rho_0) c_{jkl} \sigma_l), \quad (3.63)$$

with the partial traces resp. the reduced density matrices

$$\rho_0^s \equiv \text{Tr}_s(\rho_0), \quad s \in \{A, B\} \quad (3.64)$$

associated to the rank-1 projector

$$\rho_0 := |0\rangle \langle 0|. \quad (3.65)$$

By decomposing ρ_0^s into Hermitian orthonormal matrices

$$\rho_0^s = \frac{1}{n}(\mathbb{1} + \sum_{r=1}^{n^2-1} m_r \sigma_r), \tag{3.66}$$

we find from the traceless property of the Hermitian matrices $c_{jkl}\sigma_l$ that

$$\Omega_{[jk]}^s = \text{Tr}(\rho_0^s c_{jkl}\sigma_l) = \frac{1}{n}\text{Tr}((\mathbb{1} + \sum_{r=1}^{n^2-1} m_r \sigma_r)c_{jkl}\sigma_l)$$

$$= \frac{1}{n}\text{Tr}(c_{jkl}\sigma_l - \sum_{r=1}^{n^2-1} c_{jkl}m_r\sigma_r\sigma_l) = \frac{1}{n}\text{Tr}(\sum_{r=1}^{n^2-1} c_{jkl}m_r\sigma_r\sigma_l)$$

$$= \frac{1}{n}\sum_{r=1}^{n^2-1} c_{jkl}m_r\text{Tr}(\sigma_r\sigma_l) = \frac{2}{n}\sum_{r=1}^{n^2-1} c_{jkl}m_r\delta_{rl} = \frac{2}{n}\sum_{l=1}^{n^2-1} c_{jkl}m_l. \tag{3.67}$$

With

$$\sum_{l=1}^{n^2-1} c_{jkl}m_l\sigma_k = \sum_{l=1}^{n^2-1} [\sigma_j, m_l\sigma_l] \tag{3.68}$$

we find that

$$\sum_{l=1}^{n^2-1} c_{jkl}m_l = 0 \tag{3.69}$$

iff

$$\sum_{l=1}^{n^2-1} c_{jkl}m_l\sigma_k = 0. \tag{3.70}$$

Since the Lie algebra of $SU(n)$ is perfect, i.e.

$$[su(n), su(n)] = su(n), \tag{3.71}$$

it follows that the condition (3.70) is the case if and only if

$$m_l = 0 \text{ for all } l. \tag{3.72}$$

This implies $\Omega_{[jk]}^s = 0$ and, according to (3.66), a state having the property

$$\text{Tr}_s(\rho_0) = \frac{1}{n}\mathbb{1} \tag{3.73}$$

from which one concludes statement (b). $\square$

Corollary 3.9. *Lagrangian entanglement defines maximal entanglement.*

Proof. Since Schmidt-equivalence classes with equal distributed Schmidt-coefficients are maximal entangled according to the von Neumann entropy measure, which is the unique measure of entanglement for pure bipartite states [67], we can directly conclude from Theorem 2 the statement. □

We underline that this is a stronger statement than the one made by Bengtsson [8], where it has been concluded from dimensional arguments that maximal entangled states provide a Lagrangian submanifold. We have shown here that there are no further Lagrangian submanifolds, which provide Schmidt-equivalence classes of entangled states being less entangled than maximal.

3.3 A distance function to separable states

One of the first traditionally used measure of entanglement is the von Neumann entropy measure. Its mathematical motivation may be seen located within the implication of the so-called quantum coding theorem, which is essentially the quantum version of Shannon's classical coding theorem [72]. As indicated strongly in Theorem 3.8 and Corollary 3.9, we may discuss here whether our pure geometrical construction yields a different, perhaps more direct route to motivate a measure of entanglement, which 'evades' a digression on the proof of the quantum coding theorem.

Consider for this purpose the symmetric part of the pull-back tensor field coefficients on $U(n) \times U(n)$. Here we observed according Theorem 3.8 that

$$\rho_0 \text{ is separable} \Leftrightarrow G^{AB} = 0, \tag{3.74}$$

which recovers the traditional algebraic separability criteria related to Proposition 3.4 (e). As a matter of fact, it turns out that the block matrix G^{AB} encodes entanglement not only in a qualitative, but also in a quantitative way:

Theorem 3.10. *The trace*

$$\mathrm{Tr}((G^{AB})^T G^{AB}) \tag{3.75}$$

is related to the Euclidean distance from a pure state ρ_0 to a separable state on the real vector space of Hermitian matrices $u^*(\mathcal{H})$.

Proof. By setting

$$R := \rho_0 - \rho_0^A \otimes \rho_0^B \tag{3.76}$$

and taking into account the quantity

$$c\mathrm{Tr}((R)^\dagger R) \tag{3.77}$$

proposed in [58], one has to show that

$$\mathrm{Tr}((G^{AB})^T G^{AB}) = c\mathrm{Tr}((R)^\dagger R) \tag{3.78}$$

holds up to a normalization constant $c \in \mathbb{R}$. Without loss of generality we may use for this purpose the Bloch-representation expansions

$$\rho_0^A = \frac{1}{n}(\sigma_0 + r_j\sigma_j) \tag{3.79}$$

$$\rho_0^B = \frac{1}{n}(\sigma_0 + s_k\sigma_k) \tag{3.80}$$

$$\rho_0 \equiv \frac{1}{n^2}(\sigma_0 \otimes \sigma_0 + r_j\sigma_j \otimes \sigma_0 + s_k\sigma_0 \otimes \sigma_k + t_{jk}\sigma_a \otimes \sigma_b) \tag{3.81}$$

with

$$r_j := \mathrm{Tr}(\rho_0^A \sigma_j) = \mathrm{Tr}(\rho_0 \sigma_j \otimes \mathbb{1}) \tag{3.82}$$

$$s_k := \mathrm{Tr}(\rho_0^B \sigma_k) = \mathrm{Tr}(\rho_0 \mathbb{1} \otimes \sigma_k) \tag{3.83}$$

$$t_{jk} := \mathrm{Tr}(\rho_0 \sigma_j \otimes \sigma_k). \tag{3.84}$$

By means of the normalization $\mathrm{Tr}(\rho_0) = 1$ we have

$$G_{jk}^{AB} = \mathrm{Tr}(\rho_0 \sigma_j \otimes \sigma_k) - \mathrm{Tr}(\rho_0^A \sigma_j)\mathrm{Tr}(\rho_0^B \sigma_k) = t_{jk} - r_j s_k. \tag{3.85}$$

Starting with the left hand side of (3.78) we find then

$$\mathrm{Tr}((G^{AB})^T G^{AB}) = \sum_{j,k}^{n} G_{jk}^{AB\,2}$$

$$= \sum_{j,k}^{n} \mathrm{Tr}(\rho_0 \sigma_j \otimes \sigma_k)^2 + (\mathrm{Tr}(\rho_0^A \sigma_j)\mathrm{Tr}(\rho_0^B \sigma_k))^2 - 2\mathrm{Tr}(\rho_0 \sigma_j \otimes \sigma_k)\mathrm{Tr}(\rho_0^A \sigma_j)\mathrm{Tr}(\rho_0^B \sigma_k)$$

$$= \sum_{j,k}^{n} t_{jk}^2 + r_j^2 s_k^2 - 2t_{jk}r_j s_k. \tag{3.86}$$

By focusing on the right hand side of (3.78) we get

$$\mathrm{Tr}((R)^\dagger R) = \mathrm{Tr}(\rho_0^2) + \mathrm{Tr}(\rho_A^2 \otimes \rho_B^2) - 2\mathrm{Tr}(\rho_0(\rho_A \otimes \rho_B)). \tag{3.87}$$

By expanding each term within the trace in the Bloch-representation we find

$$\mathrm{Tr}(\rho_0^2) = \sum_{j,k}^{n} \frac{1}{n^4}(n^2 + r_j^2 n + s_k^2 n + t_{jk}^2) \tag{3.88}$$

$$\mathrm{Tr}(\rho_A^2 \otimes \rho_B^2) = \sum_{j,k}^{n} \frac{1}{n^4}(n^2 + r_j^2 n + s_k^2 n + r_j^2 s_k^2) \tag{3.89}$$

$$2\mathrm{Tr}(\rho_0(\rho_A \otimes \rho_B)) = \sum_{j,k}^{n} \frac{2}{n^4}(n^2 + r_j^2 n + s_k^2 n - t_{jk}r_j s_k). \tag{3.90}$$

Hence,
$$\mathrm{Tr}((R)^\dagger R) = \sum_{j,k}^{n} \frac{1}{n^4}(t_{jk}^2 + r_j^2 s_k^2 - 2t_{jk}r_j s_k). \tag{3.91}$$

$\square$

The quantity
$$\mathrm{Tr}((G^{AB})^T G^{AB}) \tag{3.92}$$

has been proposed by Schlienz and Mahler in [71] as an *entanglement measure* within a pure algebraic setting, basically motivated by criterion (e) in Proposition 3.4. Remarkable, they conjectured a tensorial character behind the coefficient matrix G^{AB} by proposing the name '*entanglement-tensor*' without mentioning any reference to a geometric background. Our discussion so far justified the usage of this nomenclature.

Part II
Operator-valued tensor fields

4 Operator-valued tensor fields on Lie groups

The identification of tensor fields on Lie groups and on associated orbits of pure quantum states has been achieved in the previous chapters by a pull-back of a tensor field, which has been identified on the Hilbert space. A generalization of this procedure to mixed quantum states or to infinite dimensional quantum systems may become technically not straight forward. In the following we would like to propose therefore a different approach, which admits a higher flexibility in its construction and therefore also in its possible applications within more generic quantum systems. The basic idea will be to consider a direct identification of tensor fields on a Lie group, which are representation dependent and left-invariant.

4.1 Intrinsic defined tensor fields

Given a Lie group $\mathcal{G}$, we may define for this purpose as first step an intrinsic defined tensor field on it. Here we encounter a Lie algebra-valued left-invariant 1-form[7]

$$g^{-1}dg \equiv X_j \theta^j, \tag{4.1}$$

where $\{\theta_j\}_{j \in J}$ denotes a basis of left-invariant 1-forms on $\mathcal{G}$, and $\{X_j\}_{j \in J}$ establishes a Lie algebra basis. In doing this we are considering the composition law

$$\varphi : \mathcal{G} \times \mathcal{G} \to \mathcal{G} \tag{4.2}$$

as an action of $\mathcal{G}$ on itself, where our expression dg denotes the 'differentiation' with respect to the action not respect to the 'point' which is acted upon. In this way we may consider the tensor field

$$dg^{-1} \otimes dg = -g^{-1}dg \otimes g^{-1}dg, \tag{4.3}$$

by using

$$dg^{-1} = -g^{-1}dg g^{-1}. \tag{4.4}$$

The tensor we have constructed becomes then

$$-(X_j \theta^j) \otimes (X_k \theta^k) \tag{4.5}$$

which may also be written as

$$-(X_j X_k) \theta^j \otimes \theta^j. \tag{4.6}$$

Remark 4.1. *Had we chosen $dg \otimes dg^{-1}$, we would have a similar expression in terms of right-invariant 1-forms.*

---
[7] To be precise, one should write $X_j \otimes \theta^j$ to emphasize that the values have to be taken in the Lie algebra. With an abuse of notations we set in the following $X_j \otimes \theta^j \equiv X_j \theta^j$ to keep formulas readable, resp. to avoid confusions with the tensor product symbol '$\otimes$' on the module of scalar-valued 1-forms.

4.2 Representation-dependent tensor fields

At this point we may promote this intrinsic defined tensor field in various manners to what one could call in a broader sense an 'extrinsic' defined tensor field. In short: We may consider any Lie algebra representation R of $\text{Lie}(\mathcal{G}) \equiv T_e\mathcal{G}$ and replace X_j with $R(X_j)$.
For instance, if $\mathcal{G}$ acts on an 'external' manifold $\mathcal{M}$, i.e.

$$\phi : \mathcal{G} \times \mathcal{M} \to \mathcal{M} \tag{4.7}$$

we have a canonical action obtained from taking the tangent map

$$T\phi : T\mathcal{G} \times T\mathcal{M} \to T\mathcal{M}. \tag{4.8}$$

We obtain in this way a Lie algebra homomorphism into vector fields on $\mathcal{M}$. The application of the tangent functor requires the action to be differentiable. The tangent bundle over the group manifold $T\mathcal{G}$ is the semi-direct product of $\mathcal{G}$ and its Lie algebra $\text{Lie}(\mathcal{G}) \equiv T_e\mathcal{G}$ considered as Abelian vector group with the group acting via the adjoint representation on $\text{Lie}(\mathcal{G})$. In this way we may promote the construction done in the intrinsic case to the more generic extrinsic case by starting again with an operator-valued left invariant 1-form

$$-\Phi(g)^{-1}d\Phi(g) \equiv R(X_j)\theta^j, \tag{4.9}$$

but this time being dependent on a realization

$$\Phi : \mathcal{G} \to \text{Diff}(\mathcal{M}) \tag{4.10}$$

of the Lie group action ϕ on the manifold $\mathcal{M}$ associated with the infinitesimal generator $R(X_j)$,

$$R : \text{Lie}(\mathcal{G}) \to \text{Vect}(\mathcal{M}), \tag{4.11}$$

in the module of vector fields for whatever action Φ we might consider.
An extrinsic defined tensor field on $\mathcal{G}$ is then provided by

$$-\bar{\Phi}(g)^{-1}d\Phi(g) \otimes \Phi(g)^{-1}d\Phi(g), \tag{4.12}$$

yielding

$$-R(X_j)R(X_k)\theta^j \otimes \theta^j. \tag{4.13}$$

When the homomorphism maps to vector fields, the product $R(X_j)R(X_k)$ of vector fields may associated here in various ways, either with a second order differential operator

$$L_{R(X_j)}(L_{R(X_k)}f) \tag{4.14}$$

on a function $f \in \mathcal{F}(\mathcal{M})$ or with a bi-differential operator

$$L_{R(X_j)}f_1(L_{R(X_k)}f_2) \tag{4.15}$$

on functions $(f_1, f_2) \in \mathcal{F}(\mathcal{M}) \times \mathcal{F}(\mathcal{M})$.

Now we encounter the possibility to identify $\mathcal{M}$ with several spaces of interest, including $\mathcal{G}$ itself as special case. Within the quantum kinematical settings coming either along the Schrödinger picture on a Hilbert space $\mathcal{H}$ resp. the associated space of rays $\mathcal{R}(\mathcal{H})$, or coming along the Heisenberg picture by means of the space of observables in the dual space of Hermitian matrices $u^*(\mathcal{H})$ resp. the $\mathbb{C}^*$-algebra $\mathcal{A}$ of dynamical variables, we may put particular attention on the possible identifications

$$\mathcal{M} \equiv \begin{cases} \mathcal{H} \\ \mathcal{R}(\mathcal{H}) \\ u^*(\mathcal{H}) \\ \mathcal{A}. \end{cases} \tag{4.16}$$

Unitary representations

$$U : \mathcal{G} \to U(\mathcal{H}) \tag{4.17}$$

will leave in all these cases the defining structures invariant, even though we shall distinguish between unitary vector, unitary projective and unitary adjoint representations. In this way we will find the anti-Hermitian operator-valued left-invariant 1-form

$$-U(g)^{-1}dU(g) \equiv iR(X_j)\theta^j \tag{4.18}$$

and an extrinsic defined tensor field on $\mathcal{G}$ provided by

$$-U(g)^{-1}dU(g) \otimes U(g)^{-1}dU(g), \tag{4.19}$$

yielding

$$R(X_j)R(X_k)\theta^j \otimes \theta^j, \tag{4.20}$$

if $X_j \in \text{Lie}(\mathcal{G})$ is represented with the operator $iR(X_j)$ associated with the infinitesimal unitary representation of the Lie algebra $\text{Lie}(\mathcal{G})$ acting on one of these spaces denoted by $\mathcal{M}$. Here we shall replace the product $R(X_j)R(X_k)$ with the corresponding element in the enveloping algebra of the Lie algebra[8]. Moreover, we may evaluate each of them in terms of the Hermitian product

$$\rho(R(X_j)R(X_k)) \equiv \text{Tr}(\rho\, R(X_j)R(X_k)) \tag{4.21}$$

---

[8]We recall the construction of an universal enveloping algebra $\mathcal{U}(\text{Lie}(\mathcal{G}))$ of a given Lie algebra $\text{Lie}(\mathcal{G})$. It can be given in two basic steps by the following: Take the direct sum $T(\text{Lie}(\mathcal{G})) := \oplus_n \text{Lie}(\mathcal{G})^{\otimes n}$ of all tensor product spaces $\text{Lie}(\mathcal{G})^{\otimes n}$ to build the tensor algebra $T(\text{Lie}(\mathcal{G}))$ as first step. The universal algebra is then established by considering the quotient $\mathcal{U}(\text{Lie}(\mathcal{G})) := T(\text{Lie}(\mathcal{G}))/J(\text{Lie}(\mathcal{G}))$ with $J(\text{Lie}(\mathcal{G})) \subset T(\text{Lie}(\mathcal{G}))$, an bi-literal ideal generated by all elements $A \otimes B - B \otimes A - [A,B]$ with $A, B \in \text{Lie}(\mathcal{G})$.

yielding a complex-valued tensor field

$$\rho(R(X_j)R(X_k))\theta^j \otimes \theta^j \tag{4.22}$$

on the group manifold whenever we consider a Hermitian matrix ρ in the dual $u^*(\mathcal{H})$ of the Lie algebra. In more general terms, this evaluation may be considered in analogy to the evaluation of differential operators resp. vector fields on a function $f \in \mathcal{F}(\mathcal{M})$ if we identify ρ with a linear functional

$$\rho \in \text{Lin}(\mathcal{A}) \equiv \mathcal{A}^* \tag{4.23}$$

on the C^*-algebra of complex matrices (containing $u^*(\mathcal{H})$ as real elements) for $\mathcal{M} \equiv \mathcal{A}$. The tensor field (4.22) can now be decomposed into a symmetric and an anti-symmetric part

$$\rho([R(X_j), R(X_k)]_+)\theta^j \odot \theta^j \tag{4.24}$$

$$\rho([R(X_j), R(X_k)]_-)\theta^j \wedge \theta^j. \tag{4.25}$$

whose coefficients define scalar-valued functions

$$\text{Lie}(\mathcal{G}) \times \text{Lie}(\mathcal{G}) \times \mathcal{A}^* \to \mathbb{C} \tag{4.26}$$

which become bi-linear on the Lie algebra $\text{Lie}(\mathcal{G})$ once the linear functional $\rho \in \mathcal{A}^*$ is fixed. On the other hand we are also allowed to associate with this map to any two elements in $\text{Lie}(\mathcal{G})$ a function on $\mathcal{A}^*$.

Remark 4.2. *This setting applies naturally also on Hilbert spaces $\mathcal{H}$, whenever we consider the latter as an orbit related to the action of $\mathcal{A}$ on a positive definite and normalized functional $\rho \in \mathcal{A}^*$ in the dual algebra within the GNS-construction[9].*

A general tensor field of order-k on a Lie group is defined by taking the k-th product of left-invariant 1-forms

$$-U(g)^{-1}dU(g) \otimes U(g)^{-1}dU(g) \otimes ... \otimes U(g)^{-1}dU(g), \tag{4.27}$$

yielding

$$R(X_{j_1})R(X_{j_2})..R(X_{j_k})\theta^{j_1} \otimes \theta^{j_2} \otimes .. \otimes \theta^{j_k}. \tag{4.28}$$

An evaluation of this higher rank operator-valued tensor field on a linear functional $\rho \in \mathcal{A}^*$ according to

$$\rho(R(X_{j_1})R(X_{j_2})..R(X_{j_k}))\theta^{j_1} \otimes \theta^{j_2} \otimes .. \otimes \theta^{j_k} \tag{4.29}$$

provides the map

$$\text{Lie}(\mathcal{G}) \times \text{Lie}(\mathcal{G}) \times .. \times \text{Lie}(\mathcal{G}) \times \mathcal{A}^* \to \mathbb{C} \tag{4.30}$$

for each coefficient and therefore a corresponding linear functional-dependend multi-linear form

---

[9]See also appendix in section A.

on the Lie algebra. This suggests the following definition:

Definition 4.3 (Left-invariant representation-dependent operator-valued tensor field (LIROVT)). Let $\{\theta_j\}_{j\in J}$ be a basis of left-invariant 1-forms on $\mathcal{G}$, and let $\{X_j\}_{j\in J}$ be a Lie algebra basis of $\mathcal{G}$ with $\{iR(X_j)\}_{j\in J}$, its representation in the Lie algebra $u(\mathcal{H})$ of $U(\mathcal{H})$ associated to a unitary representation $U: \mathcal{G} \to U(\mathcal{H})$. The composition

$$\left(\prod_{a=1}^{k} R(X_{j_a})\right) \bigotimes_{a=1}^{k} \theta^{j_a} \tag{4.31}$$

defines then a covariant LIROVT of order-k on the Lie group $\mathcal{G}$, associating to any group element $g \in \mathcal{G}$ a map

$$T_g\mathcal{G} \times T_g\mathcal{G}.. \times T_g\mathcal{G} \times \mathcal{A}^* \to \mathbb{C} \tag{4.32}$$

and therefore a k-multilinear form on its Lie algebra $Lie(\mathcal{G})$ for any evaluation with a linear functional $\rho \in \mathcal{A}^* \cong M_n(\mathbb{C})$ with $n < \infty$.

Remark 4.4. *This construction may be considered as generalization to Hermitian manifolds of Poincaré absolute and relative invariants within the framework of symplectic mechanics [6].*

Instead of considering tensor products of operator-valued left-invariant 1-forms $U(g)^{-1}dU(g)$, we could also apply a sequence of exterior derivatives on the latter 1-form yielding a corresponding operator-valued k-form. In the simplest case we find an operator-valued 2-form by considering

$$d(U(g)^{-1}dU(g)) = dU(g)^{-1} \wedge dU(g)$$
$$= -U(g)^{-1}dU(g) \wedge U(g)^{-1}dU(g). \tag{4.33}$$

Its evaluation with a linear functional

$$-\rho(U(g)^{-1}dU(g) \wedge U(g)^{-1}dU(g)) = \rho(R(X_j)R(X_k))\theta^j \wedge \theta^j \tag{4.34}$$

turns out to be identical with the anti-symmetrized part of the rank-2 tensor field in (4.22). This can be seen within the decomposition of the product of generators $R(X_j)R(X_k)$, where we find

$$\rho([R(X_j)R(X_k)]_+)\theta^j \wedge \theta^j + i\rho([R(X_j)R(X_k)]_-)\theta^j \wedge \theta^j,$$

with $\rho([R(X_j)R(X_k)]_+)\theta^j \wedge \theta^j = 0$.

Remark 4.5. *The 2-form*

$$\rho([R(X_j)R(X_k)]_-)\theta^j \wedge \theta^j \tag{4.35}$$

becomes non-degenerate and therefore a symplectic structure on the quotient space coming along adjoint orbits of the unitary group in the space of Hermitian matrices $u^*(\mathcal{H})$ [36].

4.3 Sums of operator-valued tensor fields

So far we considered tensor products (or exterior derivatives) on an operator-valued 1-form, being evaluated with a linear functional afterwards.

Vice versa, we may ask what happens if we consider *first* an evaluation of an operator-valued 1-form $U(g)^{-1}dU(g)$ with a linear functional according to

$$\rho(U(g)^{-1}dU(g)) \tag{4.36}$$

and *then* its tensor product

$$\rho(U(g)^{-1}dU(g)) \otimes \rho(U(g)^{-1}dU(g)) .. \otimes \rho(U(g)^{-1}dU(g)). \tag{4.37}$$

Here we will have

$$\rho(R(X_j))o(R(X_k))..\rho(R(X_l))\theta^j \odot \theta^k \odot ... \odot \theta^l \tag{4.38}$$

and therefore a symmetric tensor field. We underline that this tensor field is not identical to the symmetrization in (4.29) as one can see from the simplest case of a rank-2 tensor field

$$\rho(R(X_j))\rho(R(X_k))\theta^j \odot \theta^k, \tag{4.39}$$

which differs by means of the the anti-commutator between the Hermitian operators in

$$\rho([R(X_j), R(X_k)]_+)\theta^j \odot \theta^k. \tag{4.40}$$

Since the sum of two left-invariant tensor fields is still left invariant, we may consider in this regard both symmetric rank-2 tensor fields related into one symmetric structure by

$$(\rho([R(X_j), R(X_k)]_+) - \rho(R(X_j))\rho(R(X_k)))\theta^j \odot \theta^k. \tag{4.41}$$

Moreover, we arrive to one (but of course not the only one) possible generalization of the pullback tensor field considered in the previous chapters on pure states, by identifying the sum of left-invariant tensor fields

$$(\rho([R(X_j), R(X_k)]_+) - \rho(R(X_j))\rho(R(X_k)))\theta^j \odot \theta^k$$

$$-i\rho([R(X_j), R(X_k)]_-)\theta^j \wedge \theta^k \tag{4.42}$$

within a left-invariant tensor field. In conclusion: On a given Lie group $\mathcal{G}$ and a given Lie algebra representation R of a unitary representation in a Hilbert space $\mathcal{H}$ there exists an operator-valued tensor field $\kappa_{\mathcal{G}}$ of order two

$$\kappa_{\mathcal{G}}(\_) := (\_R(X_j)R(X_k) - \_R(X_j)\_R(X_k))\theta^j \otimes \theta^k, \tag{4.43}$$

which defines on each tangent-space over the Lie group a map

$$T_g\mathcal{G} \times T_g\mathcal{G} \times \mathcal{A}^* \to \mathbb{C}, \tag{4.44}$$

which coincides with a covariance matrix, when evaluated on a linear functional $\rho \in \mathcal{A}^*$ which is positive and normalized. This can be seen by identifying the coefficients of

$$\kappa_\mathcal{G}(\rho) := K_{jk}(\rho)\theta^j \otimes \theta^k, \tag{4.45}$$

with

$$K_{jk}(\rho) := \rho([R(X_j), R(X_k)]) - \rho(R(X_j))\rho(R(X_k)) \tag{4.46}$$

where we find the map

$$\mathrm{Lie}(\mathcal{G}) \times \mathrm{Lie}(\mathcal{G}) \times \mathcal{A}^* \to \mathbb{C} \tag{4.47}$$

$$(X_j, X_k, \rho) \mapsto K_{jk}(\rho). \tag{4.48}$$

Hence, $\kappa_\mathcal{G}$ decomposes into

$$\kappa_\mathcal{G} = G + i\Omega, \tag{4.49}$$

a symmetric part G, containing a linear part L, and an anti-symmetric part Ω according to

$$L(\rho) := \rho([R(X_j), R(X_k)]_+)\theta^j \odot \theta^k, \tag{4.50}$$

$$G(\rho) := L(\rho) - \rho(R(X_j))\rho(R(X_k))\theta^j \odot \theta^k, \tag{4.51}$$

$$\Omega(\rho) := \rho([R(X_j), R(X_k)]_-)\theta^j \wedge \theta^k. \tag{4.52}$$

It generalizes therefore the pull-back tensor field construction on Lie groups applied on pure states to operator-valued tensor fields on Lie groups applied on mixed states.

Remark 4.6. *A further generalization from finite to infinite dimensions is possible once the state ρ is considered in the domain of the infinitesimal generator $R(X_j)$, resp. in the domain of a combination $[R(X_j), R(X_k)]_\pm$ of self-adjoint operators. In particular for pure states one has to require that the associated Hilbert space vector is analytical or smooth [26, 63, 64].*

5 Application on the separability problem in composite systems $D(\mathcal{H}_A \otimes \mathcal{H}_B)$

We may now continue our discussion of section 3 on the characterization of quantum entanglement of bi-partite systems from pure to mixed states based on the framework on operator-valued tensor fields considered in the previous section. For this purpose, let us first recall the basic notions coming along the set of mixed quantum states.

5.1 The set of mixed quantum states

Within the Schrödinger picture we may describe a mixed quantum state as a convex combination

$$\rho = \sum_{j=1}^{k} p_j [\psi_j], \quad \sum_{j=1}^{k} p_j = 1, \quad p \in \mathbb{R}^+ \tag{5.1}$$

of pure quantum states $[\psi_j] \in \mathcal{R}(\mathcal{H})$. It is clear that ρ will not more be an element of $\mathcal{R}(\mathcal{H})$. To grasp the full set of mixed states in a mathematical well-defined setting, we may consider an embedding of the space of pure states into an affine or linear space. This can be done by means of a momentum map

$$\mu : \mathcal{H}_0 \to u^*(\mathcal{H}), \tag{5.2}$$

$$|\psi\rangle \mapsto |\psi\rangle \langle \psi| := \rho_\psi \tag{5.3}$$

on the real vector space $u^*(\mathcal{H})$ of Hermitian matrices. This map has the image $\mu(\mathcal{H}_0)$, given by the submanifold

$$D^1(\mathcal{H}) \subset u^*(\mathcal{H}) \tag{5.4}$$

of rank-1 Hermitian operators. If one uses the metric given by the trace, it defines a co-vector field

$$D^1(\mathcal{H}) \to u^*(\mathcal{H}) \tag{5.5}$$

$$\rho_\psi \mapsto \mathrm{Tr}(\rho_\psi, \cdot) \tag{5.6}$$

on the image $\mu(\mathcal{H}_0) = D^1(\mathcal{H})$. By restricting the momentum map on normalized vectors on the unit sphere

$$S_1(\mathcal{H}) := \{|\psi\rangle \in \mathcal{H} | \sqrt{\langle \psi | \psi \rangle} = 1\}$$

of the Hilbert space, one finds that the commutative diagram

$$\begin{array}{ccc} \mathcal{H}_0 & \xrightarrow{R_0} & S_1(\mathcal{H}) \\ {\scriptstyle C_0}\downarrow & & \downarrow{\scriptstyle \mu|_{S_1(\mathcal{H})}} \\ \mathcal{R}(\mathcal{H}) & \xrightarrow{\iota} & u^*(\mathcal{H}), \end{array}$$

defines an embedding ι of the space of rays $\mathcal{R}(\mathcal{H})$ into $u^*(\mathcal{H})$. The image

$$\iota(\mathcal{R}(\mathcal{H})) = D_1^1(\mathcal{H}) \subset D^1(\mathcal{H}) \subset u^*(\mathcal{H}) \tag{5.7}$$

becomes in this regard the subset $D_1^1(\mathcal{H})$ of rank-1 projection operators $\rho_{\psi_j} := \rho_j$ and the convex combination

$$\rho = \sum_j p_j \rho_j, \quad \sum_j p_j = 1, \quad p_j \in \mathbb{R}^+ \tag{5.8}$$

of the latter yields the space of mixed quantum states which we denote here and in the following by $D(\mathcal{H})$.

To distinguish mixed from pure states we underline:

Proposition 5.1. *For a given quantum state $\rho \in D(\mathbb{C}^n)$, following statements are equivalent:*

$$\rho \text{ is pure} \tag{a}$$

$$\rho^2 = \rho \tag{b}$$

$$\sqrt{\sum_j r_j^2} = \sqrt{\frac{n(n-1)}{2}}, \tag{c}$$

where the latter denotes the Euclidean trace norm on the traceless part of the Bloch-representation

$$\rho = \frac{1}{n}(\sigma_0 + r_j \sigma_j). \tag{5.9}$$

Proof. (a)$\Leftrightarrow$(b): Using an orthonormal basis $\{|e_j\rangle\}_{j\in I}$ in $\mathcal{H} \cong \mathbb{C}^n$, we may find for any given state $\rho \in D(\mathbb{C}^n)$ a convex combination of pure states

$$\rho = \sum_j p_j |e_j\rangle \langle e_j| \tag{5.10}$$

yielding

$$\rho^2 = \sum_{j,k} p_j p_k |e_j\rangle \langle e_j|e_k\rangle \langle e_k|$$

$$= \sum_j p_j^2 |e_j\rangle \langle e_j| = \sum_j p_j |e_j\rangle \langle e_j|, \tag{5.11}$$

iff $\rho^2 = \rho$. But this implies $\rho = |e_j\rangle \langle e_j|$. On the other hand we have for any pure state an associated normalized vector $|\psi\rangle$, such that $\rho \equiv \rho_\psi = |\psi\rangle \langle \psi|$ with $\rho_\psi^2 = |\psi\rangle \langle \psi|\psi\rangle \langle \psi| = |\psi\rangle \langle \psi|$.
(c) Given a trace-orthonormal basis $\sigma_j \in \{\sigma_j\}_{0 \leq j \leq n^2-1}$ on the space of Hermitian matrices $u^*(n)$, we may expand any such given mixed quantum state in the Bloch-representation

$$\rho = \frac{1}{n}(\sigma_0 + r_j \sigma_j). \tag{5.12}$$

Here we find
$$\rho^2 = \frac{1}{n^2}(\sigma_0 + 2r_k\sigma_k + r_j r_k \sigma_j \sigma_k, \quad (5.13)$$
and therefore
$$\mathrm{Tr}(\rho^2) = \frac{1}{n^2}(n + 2r_j r_k \delta_{jk}). \quad (5.14)$$
With $\rho^2 = \rho$ and $\mathrm{Tr}(\rho) = 1$ it follows the statement. □

On the other extreme of *maximal mixed states*, being defined by the multiple of the identity
$$\rho^* := \frac{1}{n}\mathbb{1}, \quad (5.15)$$
we find a first application of a left-invariant operator-valued tensor field (LIROVT) considered in the previous section.

Proposition 5.2. *Let*
$$\Omega(\_) := \_([R(X_j), R(X_k)]\_)\theta^j \wedge \theta^k \quad (5.16)$$
be an anti-symmetric LIROVT defined on the Lie group $U(n)$ by means of the defining representation. For a given quantum state $\rho \in D(\mathbb{C}^n)$, following statements are equivalent:

$$\rho \text{ is maximal mixed} \quad (a)$$

$$\Omega(\rho) = 0. \quad (b)$$

$$\sqrt{\sum_j r_j^2} = 0, \quad (c)$$

where the latter denotes the Euclidean trace norm on the traceless part of the Bloch-representation
$$\rho = \frac{1}{n}(\sigma_0 + r_j \sigma_j). \quad (5.17)$$

Proof. In the defining representation we have
$$\Omega(\_) = \_([\sigma_j, \sigma_k]\_)\theta^j \wedge \theta^k = \_(c_{jkl}\sigma_l)\theta^j \wedge \theta^k. \quad (5.18)$$

For a given state $\rho \in D(\mathbb{C}^n)$ we find therefore an evaluation in scalar-valued tensor-coefficients according to
$$\Omega_{[jk]}(\rho) = \mathrm{Tr}(\rho c_{jkl}\sigma_l). \quad (5.19)$$

Within the Bloch-representation
$$\rho = \frac{1}{n}(\mathbb{1} + \sum_{j=1}^{n^2-1} r_j \sigma_j), \quad (5.20)$$
we may then conclude the statement according to the argumentation starting from (5.20) in Theorem 3.8. □

5.2 The separability problem

In the case of a product Hilbert space $\mathcal{H} \equiv \mathcal{H}_A \otimes \mathcal{H}_B$ we find in particular a composite bi-partite system being defined by means of a space of mixed states $D(\mathcal{H}_A \otimes \mathcal{H}_B)$. An arbitrary element will have the form

$$\rho = \sum_j p_j \rho_j, \quad \sum_j p_j = 1, \quad p_j \in \mathbb{R}^+, \qquad (5.21)$$

with

$$\rho_j \in D_1^1(\mathcal{H}_A \otimes \mathcal{H}_B) \cong \mathcal{R}(\mathcal{H}_A \otimes \mathcal{H}_B). \qquad (5.22)$$

Each rank-1 projector is called *separable* if it can be written as

$$\rho_j = \rho_j^A \otimes \rho_j^B \qquad (5.23)$$

with $\rho_j^s \in D_1^1(\mathcal{H}_s), s \in \{A, B\}$ according to the discussion on entanglement characterization done in section 3 in the case for pure states. In this way we encounter a generalized definition on separability and entanglement given by the following:

Definition 5.3 (Separable and entangled mixed states). *A mixed state $\rho \in D(\mathcal{H}_A \otimes \mathcal{H}_B)$ is called separable if*

$$\rho = \sum_{j=1} p_j \rho_j^A \otimes \rho_j^B, \quad \sum_j p_j = 1, \quad p \in \mathbb{R}^+ \qquad (5.24)$$

with $\rho_j^s \in D_1^1(\mathcal{H}_s), s \in \{A, B\}$, otherwise entangled.

The separability problem in this generalized setting is known to be classified as a computational NP-hard problem [43]. Several approaches to this problem have been reviewed for instance in [10] by underlining the still open question whether there exists an approach which can be called *computable* in arbitrary dimensions.

5.2.1 Approach in the Bloch-representation

By starting with the algebraic approach on pure state entanglement in the Bloch-representation discussed in Proposition 3.4(e) one may find a direct extension to mixed states entanglement characterization by the following [28]:

Proposition 5.4. *A given state $\rho \in D(\mathcal{H}_A \otimes \mathcal{H}_B) \cong D(\mathbb{C}^n \otimes \mathbb{C}^n)$ in the Bloch-representation*

$$\rho \equiv \frac{1}{n^2}(\sigma_0 \otimes \sigma_0 + n_j \sigma_j \otimes \sigma_0 + m_k \sigma_0 \otimes \sigma_k + C_{jk} \sigma_j \otimes \sigma_k). \qquad (5.25)$$

with

$$n_j := \operatorname{Tr}(\rho^A \sigma_j) = \operatorname{Tr}(\rho \sigma_j \otimes \mathbb{1}) \qquad (5.26)$$

$$m_k := \operatorname{Tr}(\rho^B \sigma_k) = \operatorname{Tr}(\rho \mathbb{1} \otimes \sigma_k) \qquad (5.27)$$

$$C_{jk} := \operatorname{Tr}(\rho \sigma_j \otimes \sigma_k). \qquad (5.28)$$

is separable iff there exists a decomposition into

$$n_j = \sum_i p_i n_j^i \tag{5.29}$$

$$m_k = \sum_i p_i m_k^i \tag{5.30}$$

$$C_{jk} = \sum_i p_i n_j^i m_k^i \tag{5.31}$$

with Bloch-representation coefficients

$$n_j^i := \text{Tr}(\rho_i^A \sigma_j) \tag{5.32}$$

$$m_k^i := \text{Tr}(\rho_i^B \sigma_k). \tag{5.33}$$

associated to pure states $\rho_i^s \in D_1^1(\mathcal{H}_s), s \in \{A, B\}$.

Proof. This follows from direct computation: By considering the pure states

$$\rho_i^A = \frac{1}{n}(\sigma_0 + n_j^i \sigma_j) \tag{5.34}$$

$$\rho_i^B = \frac{1}{n}(\sigma_0 + m_k^i \sigma_k) \tag{5.35}$$

within a separable mixed state

$$\rho = \sum_i p_i \rho_i^A \otimes \rho_i^B \tag{5.36}$$

one finds

$$\rho = \frac{1}{n^2} \sum_i p_i (\sigma_0 + n_j^i \sigma_j) \otimes (\sigma_0 + m_k^i \sigma_k)$$

$$= \frac{1}{n^2} \sum_i p_i (\sigma_0 \otimes \sigma_0 + n_j^i \sigma_j \otimes \sigma_0 + m_k^i \sigma_0 \otimes \sigma_k - n_j^i m_k^i \sigma_j \otimes \sigma_k)$$

$$= \frac{1}{n^2} \left(\sum_i p_i \sigma_0 \otimes \sigma_0 + \sum_i p_i n_j^i \sigma_j \otimes \sigma_0 + \sum_i p_i m_k^i \sigma_0 \otimes \sigma_k + \sum_i p_i n_j^i m_k^i \sigma_j \otimes \sigma_k \right).$$

By comparing the latter expression with a given state $\rho \in D(\mathcal{H}_A \otimes \mathcal{H}_B)$ in the Bloch-representation (5.25) one concludes the statement. $\square$

It has been shown that this criterion implies several either sufficient or necessary separability criteria of computable nature [28]. Without proof, we restate here one of the sufficient criteria:

Proposition 5.5. *A state $\rho \in D(\mathbb{C}^n \otimes \mathbb{C}^n)$ in the Bloch-representation*

$$\rho \equiv \frac{1}{n^2}(\sigma_0 \otimes \sigma_0 + n_j \sigma_j \otimes \sigma_0 + m_k \sigma_0 \otimes \sigma_k + C_{jk} \sigma_j \otimes \sigma_k) \tag{5.37}$$

is separable if it fulfills the inequality

$$\sqrt{\frac{2(n-1)}{n}}\left(\sqrt{\sum_j n_j^2} + \sqrt{\sum_k m_k^2}\right) + \frac{2(n-1)}{n}\mathrm{Tr}(\sqrt{C^\dagger C}) \leqslant 1 \tag{5.38}$$

with a coefficient matrix $C := (C_{jk})_{j,k \in J}$.

In the following we are going to establish a link between a necessary criterion and a certain type of operator-valued tensor fields on Lie groups having been constructed in the previous section.

5.2.2 A connection to LIROVTs

In the special case of pure states entanglement we observed that a local analysis by means of tensor fields on the Lie group $U(n) \times U(n)$ and associated orbits in the composite Hilbert space provided a way to *reduce the computational effort* by avoiding a singular value decomposition into a Schmidt-basis. On the other hand, as it has been remarked in [2], the identification of a Schmidt-basis on the level of the tensor product vector space $u^*(n) \otimes u^*(n)$ implies via the restriction

$$D(\mathbb{C}^n \otimes \mathbb{C}^n) \subset u^*(n) \otimes u^*(n). \tag{5.39}$$

also in the case of mixed quantum states the possibility to extract a certain amount of information on entanglement, even though the statements become weaker than in the case of pure states.

Hence, we may again start with the set of Schmidt-decomposition inducing transformations, given by the Lie group $U(n) \times U(n)$ and consider an identification of tensor fields on corresponding orbits. This time however, we shall make use of the left-invariant representation-dependent operator-valued tensor fields (LIROVT), which have been constructed in the previous section rather then pull-back tensor fields. In this way we find a class of computable separability criteria, which are directly linked with the criteria in the Bloch-representation proposed by de Vicente [28]. It may be translated into our geometric setting by the following:

Theorem 5.6. *Let*

$$L(\_) := \_([R(X_j), R(X_k)]_+) \theta^j \odot \theta^k \tag{5.40}$$

be a symmetric LIROVT on $\mathcal{G} = U(n) \times U(n)$ *associated to a tensor product representation* $U : \mathcal{G} \to U(n^2)$. *The evaluation of the coefficients*

$$L_{(jk)}(\rho) := \mathrm{Tr}(\rho[R(X_j), R(X_k)]_+), \tag{5.41}$$

on a state $\rho \in D(\mathbb{C}^n \otimes \mathbb{C}^n)$ *implies then for*

$$L_{(jk)}|_{j,k \in J}(\rho) := C_{jk} \tag{5.42}$$

in $J := \{j,k | 1 \leq j \leq n^2 - 1, n^2 \leq k \leq 2n^2 - 2\}$ the inequality

$$\mathrm{Tr}(\sqrt{C^\dagger C}) \leq \frac{n(n-1)}{2} \tag{5.43}$$

if ρ is separable.

Proof. According to the product representation (2.147) one finds that the required coefficients are given in the block elements $L_{(jk)}|_{j,k \in J} := C_{jk}$ of the coefficient matrix $(L_{(jk)})$, defined by

$$C_{jk}(\rho) = \mathrm{Tr}(\rho[\sigma_j \otimes \mathbb{1}, \mathbb{1} \otimes \sigma_{k-n^2}]_+), \tag{5.44}$$

yielding

$$C_{jk}(\rho) = \mathrm{Tr}(\rho \sigma_j \otimes \sigma_{k-n^2}). \tag{5.45}$$

For a separable state we find by means of (5.31) in Proposition 5.4 that

$$C_{jk}(\rho) = \sum_i p_i n_j^i m_k^i \tag{5.46}$$

holds, i.e. we may identify a matrix C being decomposed into a sum

$$C \equiv \sum_{j,k} C_{jk}(\rho) \, |e_j\rangle \langle e_k| = \sum_{i,j,k} p_i n_j^i m_k^i \, |e_j\rangle \langle e_k| \tag{5.47}$$

by considering $\{|e_j\rangle\}_{j \in I}$ as canonical orthonormal basis on a real vector space. Within the Ky Fan Norm

$$\|C\|_{KF} := \mathrm{Tr}(\sqrt{C^\dagger C}), \tag{5.48}$$

we find[10]

$$\|C\|_{KF} = \|\sum_i p_i C^i\|_{KF} \leq \sum_i p_i \|C^i\|_{KF} \tag{5.49}$$

with $C^i := \sum_{j,k} n_j^i m_k^i \, |e_j\rangle \langle e_k|$. By defining the real vectors

$$|m^i\rangle := \sum_k m_k^i |e_k\rangle, \tag{5.50}$$

---

[10] Like any other matrix norm it fulfills the three fundamental properties: Positive definiteness

$$\|A\| \leq 0,$$

homogeneity

$$\|cA\| = c\|A\|$$

for scalars $c \in \mathbb{C}$, and the triangle inequality

$$\|A + B\| \leq \|A\| + \|B\|.$$

and their norm

$$|\langle m^i | m^i \rangle| := \sqrt{\sum_k (m^i_k)^2} \qquad (5.51)$$

associated to the standard Euclidean norm, which coincides with the Euclidean trace norm on the traceless part of a reduced density state, we find

$$C^i = \sum_{j,k} n^i_j m^i_k |e_j\rangle \langle e_k| = |n^i\rangle \langle m^i|$$

$$= \frac{|\langle n^i | n^i \rangle| |\langle m^i | m^i \rangle|}{|\langle n^i | n^i \rangle| |\langle m^i | m^i \rangle|} |n^i\rangle \langle m^i|$$

$$:= |\langle n^i | n^i \rangle| |\langle m^i | m^i \rangle| |\widetilde{n}^i\rangle \langle \widetilde{m}^i| \qquad (5.52)$$

with unit vectors

$$|\widetilde{n}^i\rangle := \frac{|n^i\rangle}{|\langle n^i | n^i \rangle|}, \quad |\widetilde{m}^i\rangle := \frac{|m^i\rangle}{|\langle m^i | m^i \rangle|}. \qquad (5.53)$$

This implies

$$\|C^i\|_{KF} = \text{Tr}(\sqrt{C^{i\dagger} C^i})$$

$$= |\langle n^i | n^i \rangle| |\langle m^i | m^i \rangle| \text{Tr}(\sqrt{|\widetilde{m}^i\rangle \langle \widetilde{n}^i | \widetilde{n}^i\rangle \langle \widetilde{m}^i|})$$

$$= |\langle n^i | n^i \rangle| |\langle m^i | m^i \rangle| \text{Tr}(\sqrt{|\widetilde{m}^i\rangle \langle \widetilde{m}^i|})$$

$$= |\langle n^i | n^i \rangle| |\langle m^i | m^i \rangle|$$

$$= \sqrt{\sum_j (n^i_j)^2} \sqrt{\sum_k (m^i_k)^2}$$

$$= \sqrt{\frac{n(n-1)}{2}} \sqrt{\frac{n(n-1)}{2}} \qquad (5.54)$$

according to Proposition 5.1 (c). Hence, we have with (5.49)

$$\|C\|_{KF} \leqslant \frac{n(n-1)}{2} \sum_i p_i. \qquad (5.55)$$

$\square$

Theorem 5.7. *Let*

$$\Omega(\_) := \_([R(X_j), R(X_k)]_-) \theta^j \wedge \theta^k \qquad (5.56)$$

be an anti-symmetric LIROVT on $\mathcal{G} = U(n) \times U(n)$ *associated to a tensor product representation* $U : \mathcal{G} \to U(n^2)$. *A state* $\rho \in D(\mathbb{C}^n \otimes \mathbb{C}^n)$ *with*

$$\Omega(\rho) = 0, \qquad (5.57)$$

is separable if it fulfills the inequality

$$\frac{2(n-1)}{n}\|C\|_{KF} \leq 1, \qquad (5.58)$$

with a coefficient matrix C, defined as in Theorem 5.6 by the evaluation on a symmetric LIROVT.

Proof. According to the product representation (2.147) one finds that the non-trivial coefficients are given in the block elements $\Omega_{[jk]}|_{j,k \in I} := \Omega^s_{[jk]}$ with $s \in \{A, B\}$ of the coefficient matrix $(\Omega_{[jk]})$, defined by

$$\Omega^A_{[jk]}(\rho) = \mathrm{Tr}(\rho[\sigma_j, \sigma_k]_- \otimes \mathbb{1}), \qquad (5.59)$$

$$\Omega^B_{[jk]}(\rho) = \mathrm{Tr}(\rho \mathbb{1} \otimes [\sigma_{j-n^2}, \sigma_{k-n^2}]_-). \qquad (5.60)$$

Both cases read

$$\Omega^s_{[jk]} = \mathrm{Tr}(\rho^s c_{jkl}\sigma_l), \qquad (5.61)$$

with the partial traces resp. the reduced density matrices

$$\rho^s \equiv \mathrm{Tr}_s(\rho), \quad s \in \{A, B\}. \qquad (5.62)$$

By applying Proposition 5.2 in Proposition 5.5 it follows the statement. $\square$

For bi-partite 2-level systems the inequalities in the last two theorems coincide. In this way we end up with the following conclusion:

Corollary 5.8. *The inequality*

$$\|C\|_{KF} \leq 1 \qquad (5.63)$$

becomes both a sufficient and necessary separability criterion for the case $n = 2$ *with maximal mixed subsystems* [28].

5.2.3 Example: Werner states for the case $n = 2$

Let us apply the above criteria on an explicit example. Consider for this purpose a density state in $D(\mathbb{C}^2 \otimes \mathbb{C}^2)$, defined as convex combination of a maximal entangled pure state

$$|\phi^+\rangle := \frac{1}{\sqrt{2}} \begin{pmatrix} 1 \\ 0 \end{pmatrix} \otimes \begin{pmatrix} 1 \\ 0 \end{pmatrix} + \frac{1}{\sqrt{2}} \begin{pmatrix} 0 \\ 1 \end{pmatrix} \otimes \begin{pmatrix} 0 \\ 1 \end{pmatrix}, \qquad (5.64)$$

and a maximal mixed state

$$\rho^* := \frac{1}{4}\mathbb{1}, \qquad (5.65)$$

according to

$$\rho_W := x|\phi^+\rangle\langle\phi^+| + (1-x)\rho^* \qquad (5.66)$$

with $x \in [0,1]$. The latter state is referred in the literature to the class of Werner states [78]. By evaluating

$$\rho_W \in D(\mathbb{C}^2 \otimes \mathbb{C}^2), \tag{5.67}$$

on a symmetric LIROVT

$$L(\_) := \_([R(X_j), R(X_k)]_+)\theta^j \odot \theta^k \tag{5.68}$$

associated to a product representation $SU(2) \times SU(2) \to U(4)$ we find the tensor coefficients

$$(L_{(jk)})(\rho_W) = \begin{pmatrix} 1 & 0 & 0 & x & 0 & 0 \\ 0 & 1 & 0 & 0 & -x & 0 \\ 0 & 0 & 1 & 0 & 0 & x \\ x & 0 & 0 & 1 & 0 & 0 \\ 0 & -x & 0 & 0 & 1 & 0 \\ 0 & 0 & x & 0 & 0 & 1 \end{pmatrix}, \tag{5.69}$$

defined on the 6-dimensional real Lie algebra of $SU(2) \times SU(2)$ where one encounters within a decomposition

$$(L_{(jk)}) := \begin{pmatrix} A & C \\ C & B \end{pmatrix}, \tag{5.70}$$

the block elements

$$C = \begin{pmatrix} x & 0 & 0 \\ 0 & -x & 0 \\ 0 & 0 & x \end{pmatrix}. \tag{5.71}$$

The latter is identical to the symmetric tensor coefficients $L_{(jk)}$ for $1 \leq j \leq 3$ and $5 \leq k \leq 6$. By computing the Ky Fan Norm of C one finds

$$\text{Tr}(\sqrt{C^\dagger C}) = 3x, \tag{5.72}$$

where we conclude according to Corollary 5.8 that ρ_W is separable iff

$$x \leqslant \frac{1}{3}, \tag{5.73}$$

since the evaluation of ρ_W on an anti-symmetric LIROVT

$$\Omega(\_) := \_([R(X_j), R(X_k)]_+)\theta^j \wedge \theta^k \tag{5.74}$$

associated to a product representation $SU(2) \times SU(2) \to U(4)$ yields

$$\Omega(\rho_W) = 0. \tag{5.75}$$

6 Conclusions and outlook

We come to a concluding discussion by outlining some future research programs in several directions based on the framework presented here. This gives us the opportunity to reconsider and underline some main achievements of the present work within traditional and alternative perspectives on the possible role of tensor fields both in quantum information and the foundations of quantum mechanics.

6.1 The relation between pull-back tensors and operator-valued tensors

By focusing on the infinite dimensional module of order-k tensor fields on a given manifold, we may note that the identification of 'preferred' or 'canonical' tensor fields is not possible. If the manifold caries certain additional structures however, like a group, a linear or an algebraic structure, we may always associate these structures with tensorial ones. The most prominent example in this regard may be seen provided by the 1-to-1-correspondence between left-invariant vector fields, resp. 1-forms

$$\theta^j := \text{Tr}(X_j g^{-1} dg) \tag{6.1}$$

on a Lie group and basis elements X_j of the corresponding Lie algebra. In general, the association of Lie algebra generators to left-invariant *Lie algebra valued* 1-forms

$$g^{-1} dg \equiv X_j \theta^j \tag{6.2}$$

turns out to be useful to build by means of the tensor products left-invariant *operator-valued* tensor fields of arbitrary order.

The framework of quantum mechanics suggests to consider a further ingredient into any such type of tensor field construction: It is the choice of an unitary action on a Hilbert space, which implies *representation-dependent* operator-valued tensor fields.

Crucially, by evaluating these tensor fields on a linear functional we may define a multi-linear form on each tangent space over the points of a given Lie group. By means of the left-invariance of the tensor field we may end up with a representation- and quantum state-dependent multi-linear form on the Lie algebra of the Lie group.

To 'close a circle' to the Hermitian tensor fields on which we started our discussion within the first part of the work, we may ask for a relation between pull-back tensors and operator-valued tensors.

Such a relation may come along the pull-back, induced by a projection in the $U(n)$-principal fiber bundle being embedded in the Hilbert space of Hilbert Schmidt operators as indicated by the diagram below

$$\begin{array}{ccc} GL(n,\mathbb{C}) & \xrightarrow{\iota} & \mathbb{C}^n \otimes (\mathbb{C}^n)^* \\ {\scriptstyle U(n)}\downarrow & & \downarrow{\scriptstyle \tilde{\pi}} \\ \underline{D}(\mathbb{C}^n) & \longrightarrow & D(\mathbb{C}^n), \end{array}$$

where $\underline{D}(\mathbb{C}^n)$ denotes a dense subset of $D(\mathbb{C}^n)$ realized by the states of maximal rank. These projections give instances of possible generalizations of the projections

$$\begin{array}{ccc} U(k) & \xrightarrow{f} & \mathbb{C}^n - \{0\} \\ {\scriptstyle \mathcal{G}_0^{U(1)}} \downarrow & & \downarrow \pi \\ Q & \longrightarrow & \mathcal{R}(\mathbb{C}^n) \end{array}$$

on orbits Q of pure states in the projective Hilbert space. In the lines of Uhlmann's works we may find on $\mathbb{C}^n \otimes (\mathbb{C}^n)^*$ the Bures metric being related to a metric on $D(\mathcal{H})$ in analogy to the Fubini-Study metric on $\mathcal{H}_0$ given by the pull-back of a metric covariant tensor from $\mathcal{R}(\mathcal{H})$ [77]. We leave it here as an open problem to proof, first, that the Bures metric becomes identified as a modified version of the Hermitian tensor field

$$\frac{\text{Tr}(dA^\dagger \otimes dA)}{\text{Tr}(A^\dagger A)}, \tag{6.3}$$

as a pull-back of a tensor field from $D(\mathcal{H})$, where the differential d acts on the basis-expansion coefficients of a given element A, a vector-valued 0-form on $\mathbb{C}^n \otimes (\mathbb{C}^n)^*$. And, second, that the pull-back of the Hermitian tensor field (6.3) induced by a Lie group action on a fiducial operator in $A_0 \in \mathbb{C}^n \otimes (\mathbb{C}^n)^*$, becomes related to a left-invariant operator-valued tensor field on the corresponding Lie group when evaluated on a quantum state $\tilde{\pi}(A_0) \in D(\mathbb{C}^n)$ as illustrated in the first commutative diagram.

6.2 The identification of entanglement measures

In the present work we have seen that representation- and state-dependent left-invariant covariant tensor-fields on a Lie group manifold encode certain properties of an orbit of quantum states. Being homogeneous spaces, this allowed us to deal with the idea of extracting the topological information of the latter spaces by restricting the attention on the evaluation of the tensor field coefficients on a tangent space over an arbitrary point. Hence, to detect on which topological type of orbit a quantum state 'lives', it has been sufficient to consider its local surrounding by means of left-invariant tensor fields.

These tensor fields turned out to be useful in particular for the characterization of quantum entanglement, avoiding the need of the computational effort of a singular value decomposition into Schmidt-coefficients. Our local geometric analysis illustrated this by the identification of a state as an element of a Schmidt equivalence class being covered by an orbit of local unitarily related vectors in a Hilbert space. The corresponding tensorial approach in section 3.2 implied in particular an identification of *separable, intermediate entangled and maximal* entangled pure bi-partite states and *evaded* the introduction of the von Neumann entropy, which commonly needs to be motivated by the proof of the quantum coding theorem, the analog of Shannon's coding theorem.

This has been achieved by means of product representation-dependent pull-back tensor fields of order 2 on the local unitary group of transformations (Theorem 3.8). In particular we used for this purpose a pull-back tensor field (Theorem 2.4), being identified on the corresponding Hilbert space as a structure which is a pull-back induced by the projection on the projective Hilbert space (Proposition 1.6).

In this regard it has been shown that a real sub-matrix G^{AB} composed out of the symmetric pull-back tensor coefficients exists, whose inner product

$$\text{Tr}\big((G^{AB})^T G^{AB}\big) \tag{6.4}$$

on the corresponding space of real matrices is related to an Euclidean distance between a state on the orbit and a separable state provided by the product of reduced density states (Theorem 3.10). This sub-matrix had the entries

$$G^{AB}_{jk}(\rho_\psi) = \text{Tr}(\rho_\psi \sigma_j \otimes \sigma_k) - \text{Tr}(\rho^A_\psi \sigma_j)\text{Tr}(\rho^B_\psi \sigma_j), \tag{6.5}$$

and recovered the 'algebraic' Bloch-representation separability criterion for pure states, which avoided the computational effort of a singular value decomposition into a Schmidt-basis (Proposition 3.4 (e)).

An open problem concerns the question whether the orbit classification of entangled states by means of Schmidt equivalence classes of bi-partite composite pure quantum states can be generalized to mixed quantum states within the category of smooth manifolds [2, 51, 75].

In the present work we proposed the idea to distinguish between entangled and separable mixed states by means of operator-valued tensor fields on the local unitary group $U(n) \times U(n)$. Here we identified a necessary criterion based on the sub-tensorial quantity

$$C^{AB}_{jk}(\rho) := \text{Tr}(\rho \sigma_j \otimes \sigma_k), \tag{6.6}$$

for arbitrary finite dimensions (Theorem 5.6). Sufficient criteria, also based on the Bloch-representation, are indeed available for arbitrary finite dimensions (Proposition 5.5), but it would be desirable to understand their geometric background as we displayed here in the case for the necessary criterion (Theorem 5.6). As a matter of fact, we found here a strong indication that the anti-symmetric tensor fields

$$\Omega^A_{[jk]}(\rho) = \text{Tr}(\rho[\sigma_j, \sigma_k]_- \otimes \mathbb{1}) \tag{6.7}$$

$$\Omega^B_{[jk]}(\rho) = \text{Tr}(\rho \mathbb{1} \otimes [\sigma_{j-n^2}, \sigma_{k-n^2}]_-) \tag{6.8}$$

may play a fundamental role to measure 'how far' the necessary criterion is from being also a sufficient criterion (Theorem 5.7), as we have seen for the particular case of 2-level systems with maximal mixed subsystems (Corollary 5.8).

To get into account stronger statements in higher dimensional systems we shall investigate in near future two possible strategies:

First, by means of reducible resp. irreducible representation dependent tensor fields on $SU(2) \times SU(2)$ in $U(n^2) \subset Aut(\mathbb{C}^n \otimes \mathbb{C}^n)$ with $n \geq 2$. As we illustrated here in section 2.3.1 for $SU(2)$ in $U(3) \subset Aut(\mathbb{C}^3)$, it is possible to extract certain properties on the superposition configurations of a 3-level system by means of a subgroup action of $U(3)$. This suggests to focus on the possibility to extract in a similar way a certain amount of information on entanglement in corresponding higher dimensional composite systems via the representations of the lower dimensional subgroup $SU(2) \times SU(2)$.

And second, by considering the construction of a *quantitative* geometric characterization of entanglement. In this regard we may compare (6.6) with one of the building block terms (6.5) for a quantitative description of entanglement in the case of pure states by means of the measure of Schlienz and Mahler derived in Theorem 3.10

$$f(\rho_\psi) := \mathrm{Tr}((G^{AB}(\rho_\psi))^T G^{AB}(\rho_\psi)). \tag{6.9}$$

This measure is very similar to the Ky Fan norm

$$h(\rho) := \mathrm{Tr}(\sqrt{(C^{AB}(\rho))^\dagger C^{AB}(\rho)}) \tag{6.10}$$

used in Theorem 5.6. A possible extension to a mixed states entanglement measure could be directly approached by considering the quantities

$$F(\rho) := \sum_{\inf \rho = \sum p_j \rho_j} p_j f(\rho_j) \tag{6.11}$$

$$H(\rho) := \sum_{\inf \rho = \sum p_j \rho_j} p_j h(\rho_j) \tag{6.12}$$

as discussed in [40, 41] for the case of the concurrence measure [25, 81]. Different norms may yield in this regard statements of different strength resp. methods of different computability.

In the end, we believe that a more comprehensive understanding of entanglement characterization should come along by the interplay between symmetric and anti-symmetric structures, as we have seen in Theorem 5.7. As a matter of fact, a 'unified' point of view may come along the operator-valued tensor field construction

$$\kappa_{\mathcal{G}}(\_) = (\_R(X_j)R(X_k) - \_R(X_j)\_R(X_k))\theta^j \otimes \theta^k, \tag{6.13}$$

on a given classical Lie group $\mathcal{G}$ including both symmetric and anti-symmetric tensorial structures as shown in section 4.3.

Besides recovering the class of tensor fields applied on entanglement characterization in the present work, the structure $\kappa_{\mathcal{G}}$ would reflect in this regard the 'non-commutativity' of considering first a tensor product of a left-invariant Lie algebra valued 1-form with its following evaluation with a state on the one hand, *or*, of considering first an evaluation of a state on a left-invariant Lie algebra valued 1-form with its following tensor product on the other hand:

$$
\begin{array}{ccc}
R(X_j)\theta^j & \longrightarrow & R(X_j)R(X_k)\theta^j \otimes \theta^k \\
\downarrow & & \downarrow \sharp \\
\rho(R(X_j))\theta^j & \longrightarrow & \rho(R(X_j))\rho(R(X_k))\theta^j \otimes \theta^k.
\end{array}
$$

Note that the diagram becomes commutative in certain symmetric sub-tensorial structures, as we have underlined for the case of *pure and separable* states in (6.5). Here we note that the evaluation of $\kappa_{\mathcal{G}}$ on a state coincides with the coefficients

$$K_{jk}(\rho) = \mathrm{Tr}(\rho R(X_j)R(X_k)) - \mathrm{Tr}(\rho R(X_j))\mathrm{Tr}(\rho R(X_k)), \tag{6.14}$$

of a covariance matrix, not only for pure states (Corollary 2.9), but also for mixed states. In this way we shall find a neat connection between the operator-valued tensor field $\kappa_{\mathcal{G}}$ and a recently proposed necessary separability criterion [39], known as the *covariance matrix criterion* (CMC). It is known for taking into account a whole class of both necessary and computable separability criteria for mixed quantum states available in the literature [39], including those found in the Bloch-representation by de Vicente [28]. The CMC has been anticipated in the characterization of entanglement with uncertainty relations in finite dimensional Hilbert spaces [42, 47], but may be dated back to even more older results known in systems with continuous variables: Interestingly, in particular for Gaussian states associated to the Heisenberg group one finds here a both sufficient and necessary separability criterion (see e.g. [37]).

Hence, based on the operator-valued tensor field $\kappa_{\mathcal{G}}$ one may also investigate a geometric interpretation of entanglement characterization on infinite dimensional quantum systems, whenever we take into account regularity conditions on the corresponding states (Remark 4.6). Moreover, we may consider in this regard not only Hilbert spaces associated to the Heisenberg-group resp. linear 'phase spaces', but also Hilbert spaces realized on any other Lie group $\mathcal{G}$ and related homogeneous space manifolds $\mathcal{G}/\mathcal{G}_0$.

6.3 Dirac's program revisited: The role of tensors in the foundations of quantum mechanics

Within the entanglement characterization of *pure* states we noted the remarkable fact that a 'complete break-down' of the symplectic structure, being most familiar from classical mechanics, occurs exactly if and only if we deal with maximal entangled states (Corollary 3.9). It solidifies the conjecture that no classical fundamental appearing structures should exist in that case.

In contrast to anti-symmetric structures we find that symmetric resp. Riemannian structures overcome this conjecture by establishing, as mentioned before, an essential ingredient on quantum entanglement quantification by means of the sub-tensorial structure (6.5).

We may therefore outline a picture which seems to suggest at least for *pure bi-partite* states two fundamental oppositions given by the following relations,

$$\text{Riemannian (sub-)structure} \sim \text{Quantum entanglement}$$

vs.

$$\text{symplectic structure} \sim \text{Classical separability.}$$

Such an opposition may recall Dirac's program of identifying the role of classical appearing structures in the foundations of quantum mechanics. For the purpose of generalizing and discussing these relations to the regime of *mixed bi-partite* states however, we underline that further considerations have to be taken into account.

For instance, since any given mixed state may be *purified* by coupling it to a bigger system[11], it may become more appropriated to approach the role of the symplectic structure in a *pure multi-partite* system rather then in a (corresponding partial traced and therefore reduced) mixed bi-partite system.

As a matter fact, in the latter approach we observed that a maximal degenerate anti-symmetric structure has *not* been an obstruction for the existence of separable states, as it has been been illustrated by the example of Werner states in a 2-level bi-partite system. Interestingly, indeed, these Werner states are particularly known for providing a *local hidden variable model*, even though they are entangled [78]. As it has been underlined in [11], this suggests to consider a new type of nonlocality classification scheme in dependence of the number of measurements and quantum operations needed to unveil a violation of Bell-type inequalities coming along an extension of local hidden variable models in terms of generalized observables.

To take into account the possible role of tensorial structures in the concrete formulation of such generalized resp. alternative models, we shall extend the geometric approach on entanglement characterization from finite to *infinite* dimensional Hilbert spaces realized on some 'classical' configuration space of nonlocal hidden variables.

Here we may encounter quantum nonlocality in terms of *Bohmian gradient dynamical systems*, which recover the same empirical predictions as standard quantum mechanics [34,35]. Concerning a corresponding geometric formulation, one may focus here on the unitary representations of the Heisenberg group acting on a Gaussian fiducial state to identify pull-back tensors on phase space to recover the Euclidean metric tensor for defining the gradient of Bohmian vector fields on Lagrangian subspaces, as we have shown in [4]. The pull-back procedure may therefore suggest a generalized picture of Bohmian systems from

$$\text{Euclidean Lagrangian spaces}$$

to

$$\text{Quantum state dependent Riemannian Lagrangian manifolds.}$$

---

[11]Unless it describes the quantum state of the universe.

It would be interesting to investigate, whether such a generalization might turn out useful within the specific and current developments of formulating relativistic Bohmian systems on space-time foliations being generated by a wave function [31, 33]. For this purpose we shall deal either with non-linear coherent states [57], or/and with unitary representations of the Lorentz or Poincaré group instead of the standard coherent states associated to the Heisenberg group. An extension to mixed states, finally, would suggest in particular to focus on a possible interplay between the classification of hidden variable models [11], their explicit realization by means of density states generated Bohmian vector fields, also known as *W-Bohmian Mechanics* [32], and entanglement detection based on operator-valued tensor fields on Lie groups.

In conclusion, we are convinced, that tensorial constructions will not only admit further applications in current questions of quantum entanglement characterization within quantum information and computation, but also illuminate new ways towards a closer understanding of the appearing conflict between quantum nonlocality and special relativity within the foundations of quantum mechanics in near future.

Appendices

A The GNS-construction in finite dimensions

We review here the so-called Gelfand-Naimark-Segal-construction of Hilbert spaces in the case of a given finite dimensional C^*-algebra (see also [20] [44] [45]). Consider for this purpose the following definition at the first place:

Definition A.1 (Finite dimensional C^* algebra). *A finite dimensional C^*-algebra is a Banach space $\mathcal{A}$ which is endowed with an associative product*

$$\mathcal{A} \times \mathcal{A} \to \mathcal{A} \tag{A.1}$$

$$(A, B) \mapsto A \cdot_K B, \tag{A.2}$$

and an involution operation

$$\mathcal{A} \to \mathcal{A} \tag{A.3}$$

$$A \mapsto A^*, \tag{A.4}$$

which is compatible with the Banach norm according to

$$||AA^\dagger|| = ||A||^2, \tag{A.5}$$

and coincides with the conjugate transpose

$$A^* \equiv A^\dagger \tag{A.6}$$

on complex matrices.

By the isomorphism $\mathcal{A} \cong M_n(\mathbb{C})$ to a complex matrix algebra $M_n(\mathbb{C})$ we note that the usual row-by-column matrix product

$$(A, B) \mapsto AB \tag{A.7}$$

is a special example of an associative product. In general we may deal in this regard also with alternative products not necessarily coinciding with the usual matrix product. For instance we may use

$$A \cdot_K B \equiv AKB \tag{A.8}$$

with a fixed matrix $K \in M_n(\mathbb{C})$, which satisfies

$$(A \cdot_K B) \cdot_K C = A \cdot_K (B \cdot_K C), \tag{A.9}$$

for all $A, B, C \in \mathcal{A}$.

As a first and crucial step in the GNS-construction we find that it involves the identification of

a state
$$\omega \in D(\mathcal{A}) := \{\rho \in \mathcal{A}^* | \omega(\rho\rho^\dagger) \geqslant 0, \omega(\mathbb{1}) = 1\}, \tag{A.10}$$
defined as positive, normalized linear functional on $\mathcal{A}$. Associated to such a state one constructs then an inner product
$$\langle . | . \rangle_\omega : \mathcal{A} \times \mathcal{A} \to \mathbb{C} \tag{A.11}$$
$$\langle A | B \rangle_\omega := \omega(A^\dagger B) = \text{Tr}(\omega A^\dagger B), \tag{A.12}$$
which turns out to be degenerate on $\mathcal{A}$, but non-degenerate on the quotient $\mathcal{A}/\mathcal{J}_\omega$, where $\mathcal{J}_\omega$ denotes the so-called Gelfand bilateral ideal containing all $A \in \mathcal{A}$ with $\omega(A^\dagger A) = 0$. This quotient establishes a Hilbert space
$$\mathcal{A}/\mathcal{J}_\omega := \mathcal{H}_\omega \tag{A.13}$$
of equivalence classes
$$|A\rangle := [A + \mathcal{J}_\omega] \tag{A.14}$$
$$|B\rangle := [B + \mathcal{J}_\omega], \tag{A.15}$$
whenever the induced inner product $\langle A | B \rangle_\omega$ defines a norm under which $\mathcal{H}_\omega$ becomes complete. Any vector $|B\rangle$ in $\mathcal{H}_\omega$ can be obtained in this regard from a cyclic vector related to the equivalence class represented by the unit element in $\mathcal{A}$
$$|\Omega\rangle := [\mathbb{1} + \mathcal{J}_\omega] \tag{A.16}$$
according to
$$|B\rangle = \pi_\omega(B) |\Omega\rangle \in \mathcal{H}_\omega, \tag{A.17}$$
where $\pi_\omega(B)$ provides a representation of $\mathcal{A}$ on $\mathcal{H}_\omega$,
$$\pi_\omega : \mathcal{A} \to gl(\mathcal{H}_\omega). \tag{A.18}$$
In particular the vector $|\Omega\rangle \in \mathcal{H}_\omega$ becomes related to the state $\omega \in D(\mathcal{A})$ according to
$$\omega(A) = \text{Tr}(|\Omega\rangle \langle\Omega| \pi_\omega(A)). \tag{A.19}$$
More general, by using (A.17), *any* vector $|B\rangle \in \mathcal{H}_\omega$ becomes then related to a state $\omega_{|B\rangle} \in D(\mathcal{A})$
$$\omega_{|B\rangle}(A) = \text{Tr}(|B\rangle \langle B| \pi_\omega(A)). \tag{A.20}$$
This relation is of particular importance, since it implies the momentum map within the 'sequence'
$$\mathcal{H}_\omega \to u^*(\mathcal{H}_\omega) \to D(\mathcal{A}) \tag{A.21}$$
$$|B\rangle \mapsto |B\rangle \langle B| \mapsto \omega_{|B\rangle}. \tag{A.22}$$
and therefore a relation between Hilbert space vectors $|B\rangle$ in the Schrödinger picture and states

$\omega_{|B\rangle}$ in the C^*-algebraic Heisenberg picture. More explicitly we have

$$\omega_{|B\rangle}(A) = \text{Tr}(\pi_\omega(B)|\Omega\rangle\langle\Omega|\pi_\omega(B^\dagger)\pi_\omega(A))$$

$$= \text{Tr}(|\Omega\rangle\langle\Omega|\pi_\omega(B^\dagger)\pi_\omega(A)\pi_\omega(B))$$

$$= \text{Tr}(|\Omega\rangle\langle\Omega|\pi_\omega(B^\dagger AB)) =_{(A.19)} \omega(B^\dagger AB), \quad (A.23)$$

yielding
$$\text{Tr}(B\omega B^\dagger A), \quad (A.24)$$

an adjoint action of $\mathcal{A}$ on $\omega \in \mathcal{A}^*$ for all $B \in \mathcal{A}$ and therefore an orbit $\mathcal{A}\cdot\omega$ in $\mathcal{A}^*$ with

$$\mathcal{H}_\omega \equiv \mathcal{A}\cdot\omega. \quad (A.25)$$

Now we have to distinguish whether the orbit $\mathcal{H}_\omega \subset \mathcal{A}^*$ provides an irreducible or reducible representation Hilbert space. The latter case implies by means of a generalized momentum map

$$\mathcal{H}_\omega := \bigoplus_j \mathcal{H}_{\omega_j} \to u^*(\mathcal{H}_\omega) \to D(\mathcal{A}) \quad (A.26)$$

$$|\Omega\rangle := \oplus_j |\Omega_j\rangle \mapsto \sum_j p_j |\Omega_j\rangle\langle\Omega_j| \mapsto \omega_{|\Omega\rangle}. \quad (A.27)$$

a convex sum combination of pure states $|\Omega_j\rangle\langle\Omega_j|$ with

$$\sum_j p_j \equiv \sum_j \langle\Omega_j|\Omega_j\rangle = 1, \quad (A.28)$$

associated to a direct sum of irreducible representations. In particular with (A.19) and (A.20) we have

$$\omega_{\oplus_j|\Omega_j\rangle}(A) = \sum_j \text{Tr}(p_j|\Omega_j\rangle\langle\Omega_j|\pi_\omega(A)), \quad (A.29)$$

with an orbit of mixed states given by

$$\omega_{\oplus_j|B_j\rangle}(A) = \sum_j \text{Tr}(p_j\pi_\omega(B)|\Omega_j\rangle\langle\Omega_j|\pi_\omega(B^\dagger)\pi_\omega(A)). \quad (A.30)$$

Within (A.19) we may rewrite the inner product in terms of the GNS-representation according to

$$\omega(A^\dagger A) = \text{Tr}(|\Omega\rangle\langle\Omega|\pi_\omega(A^\dagger A))$$

$$= \text{Tr}(|\Omega\rangle\langle\Omega|\pi_\omega(A^\dagger)\pi_\omega(A)). \quad (A.31)$$

Hence, for performing explicit computations one has to take into account that $\langle A|A\rangle_\omega$ in (A.12) depends on a state $|\Omega\rangle\langle\Omega|$ and on a corresponding GNS representation π_ω. Of course in the defining representation one has $\pi_\omega(A) = A$, implying $\omega = |\Omega\rangle\langle\Omega|$. This in contrast to a reducible representation $\pi_\omega(A) = 1_n \otimes A$ associated to a m-rank projector $\omega = \sum_j^m p_j|\Omega_j\rangle\langle\Omega_j|$, where $\mathbb{1}_m$ denotes a $m \times m$ identity matrix. In conclusion, we will be allowed to identify *mixed states* if and only if the representation induced by ω is reducible. To see how the GNS-construction works let us focus on two explicit examples.

A.1 Hilbert spaces from pure states

Consider the expansion of an arbitrary element in $\mathcal{A}$ according to

$$A = \sum_{j,k} A_{jk} |j\rangle \langle k|, \tag{A.32}$$

where $\{|j\rangle \langle k|\}_{j,k \in I}$ denotes a basis on the complex vector space $\mathcal{A} \cong \mathbb{C}^{n^2}$. Moreover, let ω a rank-1 projection operator yielding a pure state in terms of the above basis according to

$$\omega := |1\rangle \langle 1|. \tag{A.33}$$

The inner product on $\mathcal{A}$ becomes then

$$\langle A | A \rangle_\omega = \text{Tr}(\omega A^\dagger A)$$

$$= \sum_{j,k} \sum_{l,s} \text{Tr}(|1\rangle \langle 1| \bar{A}_{kj} |j\rangle \langle k| A_{ls} |l\rangle \langle s|)$$

$$= \sum_{j,k} \sum_{l,s} \text{Tr}(|1\rangle \langle 1|j\rangle \langle k|l\rangle \langle s|) \bar{A}_{kj} A_{ls}$$

$$= \sum_{j,k} \sum_{l,s} \delta_{1j} \delta_{kl} \delta_{s1} \bar{A}_{kj} A_{ls}$$

$$= \sum_k \bar{A}_{k1} A_{k1} \equiv \sum_k \bar{a}_k a_k \tag{A.34}$$

by setting $A_{k1} \equiv a_k$. This defines a Hermitian inner product, turning the quotient space

$$\mathcal{A}/\mathcal{J}_\omega = \mathbb{C}^n \tag{A.35}$$

with the Gelfand ideal defined by

$$\mathcal{J}_\omega = \{A \in \mathcal{A} | A_{k1} = 0, k = 1, .., n\}, \tag{A.36}$$

into a Hilbert space. The latter arises then as an orbit $\mathcal{A} \cdot \omega$ in $\mathcal{A}^*$ by means of the action on $|1\rangle$ in the defining representation of $\mathcal{A}$ in $gl(\mathcal{H}_\omega)$ according to

$$\pi_\omega(A) |\Omega\rangle = A |1\rangle = \sum_{j,k} A_{jk} |j\rangle \langle k |1\rangle = \sum_j A_{j1} |j\rangle = \sum_j a_j |j\rangle \equiv |A\rangle. \tag{A.37}$$

A.2 Hilbert spaces from mixed states

Again we may consider $\{|j\rangle\langle k|\}_{j,k \in I}$ as a basis on the complex vector space $\mathcal{A} \cong \mathbb{C}^{n^2}$, but now by considering as state a rank-m projection operator yielding a *mixed* state

$$\omega := \sum_i p_i |i\rangle\langle i| \tag{A.38}$$

with $\sum_j p_j = 1$ and $p_i > 0$. In this case the inner product on $\mathcal{A}$ becomes

$$\langle A | A \rangle_\omega = \text{Tr}(\omega A^\dagger A)$$

$$= \sum_{i,j,k,l,s} \text{Tr}(p_i |i\rangle\langle i| \bar{A}_{kj} |j\rangle\langle k| A_{ls} |l\rangle\langle s|)$$

$$= \sum_{i,j,k,l,s} \text{Tr}(p_i |i\rangle\langle i|j\rangle\langle k|l\rangle\langle s|) \bar{A}_{kj} A_{ls}$$

$$= \sum_{i,j,k,l,s} \delta_{ij}\delta_{kl}\delta_{is} p_i \bar{A}_{kj} A_{ls}$$

$$= \sum_{i,k} p_i \bar{A}_{ki} A_{ki} \equiv \sum_{i,k} \bar{a}_k^{(i)} a_k^{(i)}, \tag{A.39}$$

with $a_k^{(i)} \equiv \sqrt{p_i} A_{ki}$. The corresponding Gelfand ideals reads

$$\mathcal{J}_\omega = \{A \in \mathcal{A} | A_{ki} = 0, k = 1,..,n; i = 1,..,m\}, \tag{A.40}$$

which becomes trivial for $m = n$. This defines a Hermitian inner product on an orbit $\mathcal{A} \cdot \omega$ defined in a reducible representation $\pi_\omega(A) = \mathbb{1} \otimes A$ according to

$$\pi_\omega(A) \, |\Omega\rangle = (\mathbb{1} \otimes A) \bigoplus_i^m |i\rangle = \bigoplus_i^m A |i\rangle$$

$$\bigoplus_i^m \sum_{j,k} A_{jk} |j\rangle\langle k|i\rangle = \bigoplus_i^m \sum_j A_{ji} |j\rangle = \bigoplus_i^m \sum_j \frac{a_j^{(i)}}{\sqrt{p_i}} |j\rangle \equiv |A\rangle, \tag{A.41}$$

yielding as Hilbert space the direct sum

$$\mathcal{H}_\omega = \bigoplus^m \mathbb{C}^n. \tag{A.42}$$

B Special morphisms: From embeddings to tensor products

The aim of this section is to give a small collection of definitions on some of the most frequently used mathematical notions within the present work.

Let $\mathcal{M}, \mathcal{N}$ be topological spaces with $\mathrm{Dim}(\mathcal{M}) \leqslant \mathrm{Dim}(\mathcal{N})$ and

$$f : \mathcal{M} \to \mathcal{N} \tag{B.1}$$

a morphism between them. Within the category of differentiable manifolds, we may restrict in the following on differentiable maps as morphisms. The morphism f is then called an

- *embedding* if it is an injection, which is an inclusion of the image

$$f(\mathcal{M}) \subset \mathcal{N}, \tag{B.2}$$

such that $f(\mathcal{M})$ is a submanifold of $\mathcal{N}$.

- *immersion* if it is locally an embedding, i.e. there exists to any point $m \in \mathcal{M}$ an open neighborhood $U_m \subset \mathcal{M}$, such that the restriction of f on U_m is injective and the image

$$f(U_m) \subset \mathcal{N} \tag{B.3}$$

is a submanifold of $\mathcal{N}$.

As one can illustrate in the case of an immersion of $\mathcal{M} = \mathbb{R}$ with self-intersections in $\mathcal{N} = \mathbb{R}^2$, an immersion is in general not an embedding.

Let us remark that the dual notion to immersions is given by *submersions*, which may be described in the category of manifolds, by morphisms, which admit in any local chart a projection. Hence, any projection is an instance of a submersion.

As criteria for these notions we may use the tangent map

$$df : T\mathcal{M} \to T\mathcal{N} \tag{B.4}$$

on each tangent space according to

$$df|_m : T_m\mathcal{M} \to T_{f(m)}\mathcal{N}, \tag{B.5}$$

which may be described in a local chart by a Jacobian matrix.

Here we find the following equivalent statements:

- f is an immersion

- $df|_m$ is injective for all points $m \in \mathcal{M}$.

- $\mathrm{rank}(df|_m) = \mathrm{Dim}(\mathcal{M})$ for all points $m \in \mathcal{M}$.

A morphism fails to be an embedding if one of the above requirements is not fulfilled. On the other hand we have the following equivalent statements:

- f is a submersion
- $df|_m$ is surjective for all points $m \in \mathcal{M}$
- All points $m \in \mathcal{M}$ are *regular points* of f.
- $\mathrm{rank}(df|_m) = \mathrm{Dim}(\mathcal{N})$ for all points $m \in \mathcal{M}$.

A *value* $f(m) := n$ is called in this regard *regular* if $f^{-1}(n)$ is a submanifold of regular points in $\mathcal{M}$. A particular class of submersions may come along fiber bundle projections.

The most important examples within this underlying work are the submersions of the type

$$f : \mathcal{G} \to \mathcal{G}/\mathcal{G}_0, \tag{B.6}$$

$$g \mapsto [g] \tag{B.7}$$

mapping a Lie group $\mathcal{G}$ to a homogeneous space $\mathcal{G}/\mathcal{G}_0$ with isotropy group $\mathcal{G}_0$. In this case we would have a submanifold of regular points

$$f^{-1}([g]) \cong \mathcal{G}_0, \tag{B.8}$$

being isomorphic to the closed Lie subgroup $\mathcal{G}_0 \subset \mathcal{G}$. In this regard we may observe that the submersion f may be seen induced along a 'functor' from the category of Lie groups (the intersection between manifold and group categories) to the bigger category of manifolds being not necessarily endowed with a group structure.

The other direction from a bigger to a smaller category may come along a morphism which is an embedding. Within the underlying work we encounter here in particular the embedding of orbits

$$\mathcal{G}/\mathcal{G}_0 \hookrightarrow \mathcal{H} \tag{B.9}$$

into a Hilbert space. By promoting the Hilbert space to a Hilbert manifold it becomes clear that such an embedding implies an instance of a restriction from the category of manifolds to a smaller category, in this case given by the subcategory of manifolds which are endowed with a Hermitian structure. One may nevertheless 'remain' within the same category by considering more specific embeddings, like *Hermitian embeddings*, whenever $\mathcal{G}/\mathcal{G}_0$ admits a Hermitian structure coinciding with the pull-back of the Hermitian structure available on the Hilbert manifold.

By starting with the projective Hilbert space we may mention in this regard a more particular type of embedding, which is linked to the notion of a *tensor product*. It is the so-called *Segre embedding*

$$\mathrm{Seg} : \mathcal{R}(\mathcal{H}_A) \times \mathcal{R}(\mathcal{H}_B) \hookrightarrow \mathcal{R}(\mathcal{H}_A \otimes \mathcal{H}_B) \tag{B.10}$$

providing an isomorphism between the cartesian product of two projective Hilbert spaces and the submanifold
$$\mathrm{Seg}(\mathcal{R}(\mathcal{H}_A) \times \mathcal{R}(\mathcal{H}_B)) \subset \mathcal{R}(\mathcal{H}_A \otimes \mathcal{H}_B) \tag{B.11}$$
of separable pure bi-partite states.

To motivate the idea behind the notion of a tensor product in more general terms we may start by considering a product
$$B(u,v) := u \cdot_B v \in W \tag{B.12}$$
between elements of two vector spaces, say $u \in U$ and $v \in V$ into a third vector space W over a field K. Such a product shall be defined by a bi-linear map
$$B : U \times V \to W, \tag{B.13}$$
sharing by definition the properties
$$B(u, v+w) = B(u,v) + B(u,w) \tag{B.14}$$
$$B(u+w, v) = B(u,v) + B(w,v) \tag{B.15}$$
$$B(u, rv) = B(ru, v) = rB(u,v) \tag{B.16}$$
for $r \in K$. The basic idea behind the tensor product is now to establish an universal product from which this or any other bi-linear product on $U \times V$ may be re-constructed: The tensor product between two vector spaces is defined by a bi-linear embedding
$$\bigotimes : U \times V \to U \otimes V \tag{B.17}$$
into a vector space $U \otimes V$, such that there exists a unique linear map
$$T : U \otimes V \to W \tag{B.18}$$
which factorizes a given bi-linear map into
$$B = T \circ \bigotimes . \tag{B.19}$$
In short:
$$T(u \otimes v) = u \cdot_B v. \tag{B.20}$$

References

[1] M. C. ABBATI, R. CIRELLI, P. LANZAVECCHIA, AND A. MANIÁ, *Pure states of general quantum-mechanical systems as Kähler bundles*, Nuovo Cimento B Serie, 83 (1984), pp. 43–60.

[2] P. ANIELLO AND C. LUPO, *On the relation between Schmidt coefficients and entanglement*, Open Sys. Information Dyn., 16 (2009), p. 127.

[3] P. ANIELLO, G. MARMO, J. CLEMENTE-GALLARDO, AND G. F. VOLKERT, *Classical tensors and quantum entanglement I: Pure states*, To be published in Int. J. Geom. Meth. Mod. Phys., (2010).

[4] P. ANIELLO, G. MARMO, AND G. F. VOLKERT, *Classical tensors from quantum states*, Int. J. Geom. Meth. Mod. Phys., 06 (2009), pp. 369–383.

[5] F. T. ARECCHI, E. COURTENS, R. GILMORE, AND H. THOMAS, *Atomic coherent states in quantum optics*, Phys. Rev. A, 6 (1972), pp. 2211–2237.

[6] V. I. ARNOLD, *Les methodes mathematiques de la Mecanique Classique*, Editions Mir, Moscow, (1976).

[7] A. ASHTEKAR AND T. A. SCHILLING, *Geometrical Formulation of Quantum Mechanics*, in On Einstein's Path: Essays in honor of Engelbert Schucking, A. Harvey, ed., (1999), p. 23.

[8] I. BENGTSSON, *A Curious Geometrical Fact about Entanglement*, in Quantum Theory: Reconsideration of Foundations, G. Adenier, A. Y. Khrennikov, P. Lahti, & V. I. Man'ko, ed., vol. 962 of American Institute of Physics Conference Series, (2007), pp. 34–38.

[9] I. BENGTSSON, J. BRÄNNLUND, AND K. ŻYCZKOWSKI, CP^n, *OR, Entanglement Illustrated*, International Journal of Modern Physics A, 17 (2002), pp. 4675–4695.

[10] I. BENGTSSON AND K. ŻYCZKOWSKI, *Geometry of Quantum States*, Cambridge University Press, New York, (2006).

[11] K. BERNDL, D. DÜRR, S. GOLDSTEIN, S. TEUFEL, AND N. ZANGHÌ, *Locality and causality in hidden-variables models of quantum theory*, Phys. Rev. A, 56 (1997), pp. 1217–1227.

[12] D. C. BRODY AND L. P. HUGHSTON, *Geometric quantum mechanics*, Journal of Geometry and Physics, 38 (2001), pp. 19–53.

[13] V. CANTONI, *Generalized 'transition probability'*, Communications in Mathematical Physics, 44 (1975), pp. 125–128.

[14] ——, *Intrinsic geometry of the quantum-mechanical phase space, Hamiltonian systems and Correspondence Principle*, Rend. Accad. Naz. Lincei, 62 (1977), pp. 628–636.

[15] ——, *The riemannian structure on the states of quantum-like systems*, Communications in Mathematical Physics, (1977).

[16] ——, *Geometric aspects of Quantum Systems*, Rend. sem. Mat. Fis. Milano, 48 (1980), pp. 35–42.

[17] ——, *Superposition of physical states: a metric viewpoint*, Helv. Phys. Acta, 58 (1985), pp. 956–968.

[18] J. F. CARIÑENA, J. CLEMENTE-GALLARDO, AND G. MARMO, *Geometrization of quantum mechanics*, Theoretical and Mathematical Physics, 152 (2007), pp. 894–903.

[19] J. F. CARIÑENA, J. GRABOWSKI, AND G. MARMO, *Quantum Bi-Hamiltonian Systems*, International Journal of Modern Physics A, 15 (2000), pp. 4797–4810.

[20] D. CHRUSCINSKI AND G. MARMO, *Remarks on the GNS Representation and the Geometry of Quantum States*, Open Syst. Info. Dyn., 16 (2009), pp. 157–177.

[21] R. CIRELLI AND P. LANZAVECCHIA, *Hamiltonian vector fields in quantum mechanics*, Nuovo Cimento B Serie, 79 (1984), pp. 271–283.

[22] R. CIRELLI, P. LANZAVECCHIA, AND A. MANIA, *Normal pure states of the von Neumann algebra of bounded operators as Kähler manifold*, Journal of Physics A Mathematical General, 16 (1983), pp. 3829–3835.

[23] J. CLEMENTE-GALLARDO AND G. MARMO, *The space of density states in geometrical quantum mechanics*, in 'Differential Geometric Methods in Mechanics and Field Theory', Volume in Honour of Willy Sarlet, Gent Academia Press, F. Cantrijn, M. Crampin and B. Langerock, ed., (2007), pp. 35–56.

[24] ——, *Basics of Quantum Mechanics, Geometrization and Some Applications to Quantum Information*, International Journal of Geometric Methods in Modern Physics, 5 (2008), p. 989.

[25] V. COFFMAN, J. KUNDU, AND W. K. WOOTTERS, *Distributed entanglement*, Phys. Rev. A, 61 (2000), p. 052306.

[26] E. B. DAVIES, *Hilbert space representations of Lie algebras*, Communications in Mathematical Physics, 23 (1971), pp. 159–168.

[27] M. DE GOSSON, *The principles of Newtonian and quantum mechanics*, Imperial College Press, London, (2001).

[28] J. I. DE VICENTE, *Separability criteria based on the Bloch representation of density matrices*, Quantum Inf. Comput., 7 (2007), p. 624.

[29] P. A. M. DIRAC, *On the analogy between classical and quantum mechanics*, Rev. Mod. Phys., 17 (1945), pp. 195–199.

[30] ——, *The principles of quantum mechanics*, The International Series of Monographs on Physics, Clarendon Press, Oxford, (1947).

[31] D. DÜRR, S. GOLDSTEIN, K. MÜNCH-BERNDL, AND N. ZANGHÌ, *Hypersurface Bohm-Dirac models*, Phys. Rev. A, 60 (1999), pp. 2729–2736.

[32] D. DÜRR, S. GOLDSTEIN, R. TUMULKA, AND N. ZANGHÍ, *On the Role of Density Matrices in Bohmian Mechanics*, Foundations of Physics, 35 (2005), pp. 449–467.

[33] D. DÜRR, S. GOLDSTEIN, AND N. ZANGHÌ, *On a Realistic Theory for Quantum Physics*, Stochastic Process, Physics and Geometry, edited by S. Albeverio, G. Casati, U. Cattaneo, D. Merlini, and R. Moresi, World Scientific, (1992), pp. 374–391.

[34] ——, *Quantum equilibrium and the origin of absolute uncertainty*, Journal of Statistical Physics, 67 (1992), pp. 843–907.

[35] ——, *Quantum Equilibrium and the Role of Operators as Observables in Quantum Theory*, Journal of Statistical Physics, 116 (2004), pp. 959–1055.

[36] E. ERCOLESSI, G. MARMO, G. MORANDI, AND N. MUKUNDA, *Geometry of Mixed States and Degeneracy Structure of Geometric Phases for Multi-Level Quantum Systems*, International Journal of Modern Physics A, 16 (2001), pp. 5007–5032.

[37] A. FERRARO, S. OLIVARES, AND M. G. A. PARIS, *Gaussian states in continuous variable quantum information*, Napoli Series on physics and Astrophysics, Bibliopolis Naples, (2005).

[38] G. GIBBONS, *Typical states and density matrices*, Journal of Geometry and Physics, 8 (1992), pp. 147–162.

[39] O. GITTSOVICH, O. GÜHNE, P. HYLLUS, AND J. EISERT, *Unifying several separability conditions using the covariance matrix criterion*, Phys. Rev. A, 78 (2008), p. 052319.

[40] J. GRABOWSKI, M. KUŚ, AND G. MARMO, *Geometry of quantum systems: density states and entanglement*, Journal of Physics A Mathematical General, 38 (2005), pp. 10217–10244.

[41] J. GRABOWSKI, M. KUŚ, AND G. MARMO, *Symmetries, group actions, and entanglement*, Open Sys. Information Dyn., 13 (2006), pp. 343–362.

[42] O. GÜHNE, *Characterizing Entanglement via Uncertainty Relations*, Phys. Rev. Lett., 92 (2004), p. 117903.

[43] L. GURVITS, *Classical deterministic complexity of Edmonds' Problem and quantum entanglement*, in STOC '03: Proceedings of the thirty-fifth annual ACM symposium on Theory of computing, ACM, New York, (2003), pp. 10–19.

[44] R. HAAG, *Local quantum physics: Fields, particles, algebras*, Texts and monographs in physics, Springer, Berlin, (1992), p. 356.

[45] R. HAAG AND D. KASTLER, *An Algebraic Approach to Quantum Field Theory*, Journal of Mathematical Physics, 5 (1964), pp. 848–861.

[46] A. HESLOT, *Quantum mechanics as a classical theory*, Phys. Rev. D, 31 (1985), pp. 1341–1348.

[47] H. F. HOFMANN AND S. TAKEUCHI, *Violation of local uncertainty relations as a signature of entanglement*, Phys. Rev. A, 68 (2003), p. 032103.

[48] J. R. KLAUDER, *Quantization Without Quantization*, Annals of Physics, 237 (1995), pp. 147–160.

[49] ——, *Understanding quantization*, Foundations of Physics, 27 (1997), pp. 1467–1483.

[50] ——, *Phase Space Geometry in Classical and Quantum Mechanics*, Proceedings of the Second International Workshop on Contemporary Problems in Mathematical Physics, eds. J. Govaerts, M. N. Hounkonnou and A. Z. Msezane, World Scientific, Singapore, (2002), pp. 395–408.

[51] M. KUŚ AND K. ŻYCZKOWSKI, *Geometry of entangled states*, Phys. Rev. A, 63 (2001), p. 032307.

[52] G. W. MACKEY, *Quantum Mechanics and Hilbert Space*, The American Mathematical Monthly, 64 (1957), pp. 45–57.

[53] ——, *Mathematical foundations of quantum mechanics*, Mathematical Physics Monograph series, Benjamin Cummings, New York, (1963).

[54] ——, *Weyl's program and modern physics*, Proceedings of the 16th international conference on differential geometric methods in theoretical physics, Como, Kluwer Acad. Publ., Dordrecht, (1988), pp. 11–36.

[55] ——, *The Relationship Between Classical mechanics and Quantum Mechanics*, Contem. Math., (1998), pp. 11–36.

[56] E. MAJORANA, *Atomi orientati in campo magnetico variabile*, Il Nuovo Cimento, 9 (1932), pp. 43–50.

[57] V. I. MAN'KO, G. MARMO, E. C. G. SUDARSHAN, AND F. ZACCARIA, *f-Oscillators and nonlinear coherent states*, Physica Scripta, 55 (1997), p. 528.

[58] ——, *Interference and entanglement: an intrinsic approach*, Journal of Physics A Mathematical General, 35 (2002), pp. 7137–7157.

[59] G. MARMO, G. MORANDI, A. SIMONI, AND F. VENTRIGLIA, *Alternative structures and bi-Hamiltonian systems*, Journal of Physics A Mathematical General, 35 (2002), pp. 8393–8406.

[60] D. MONTGOMERY AND L. ZIPPIN, *Topological transformation groups*, Interscience Publishers, Inc., New York, (1955).

[61] A. MOROIANU, *Lectures on Kähler Geometry*, London Mathematical Society Student Texts 69, Cambridge, (2007).

[62] H. NAKAJIMA, *Lectures on Hilbert schemes of points on surfaces*, AMS Univ. Lecture Series, 18 (1999).

[63] E. NELSON, *Analytic Vectors*, The Annals of Mathematics, Second Series, 70 (1959), pp. 572–615.

[64] E. NELSON AND W. F. STINESPRING, *Representation of elliptic operators in an enveloping algebra*, American Journal of Mathematics, 81 (1959), pp. 547–560.

[65] E. ONOFRI, *A note on coherent state representations of Lie groups*, Journal of Mathematical Physics, 16 (1975), pp. 1087–1089.

[66] A. M. PERELOMOV, *Coherent states for arbitrary Lie group*, Communications in Mathematical Physics, 26 (1972), p. 222.

[67] S. POPESCU AND D. ROHRLICH, *Thermodynamics and the measure of entanglement*, Phys. Rev. A, 56 (1997), pp. 3319–3321.

[68] J. P. PROVOST AND G. VALLEE, *Riemannian structure on manifolds of quantum states*, Communications in Mathematical Physics, 76 (1980), pp. 289–301.

[69] R. RESTA, *Electron Localization in the Quantum Hall Regime*, Physical Review Letters, 95 (2005), p. 196805.

[70] D. J. ROWE, A. RYMAN, AND G. ROSENSTEEL, *Many-body quantum mechanics as a symplectic dynamical system*, Phys. Rev. A, 22 (1980), pp. 2362–2373.

[71] J. SCHLIENZ AND G. MAHLER, *Description of entanglement*, Phys. Rev. A, 52 (1995), pp. 4396–4404.

[72] B. SCHUMACHER, *Quantum coding*, Phys. Rev. A, 51 (1995), pp. 2738–2747.

[73] I. E. SEGAL, *Postulates for General Quantum Mechanics*, The Annals of Mathematics, Second Series, 48 (1947), pp. 930–948.

[74] P. W. SHOR, *Quantum computing*, Doc. Math. J. DMV, Extra Volume ICM, 23 (1998), pp. 467–486.

[75] M. M. SINOLECKA, K. ŻYCZKOWSKI, AND M. KUŚ, *Manifolds of Equal Entanglement for Composite Quantum Systems*, Acta Physica Polonica B, 33 (2002), p. 2081.

[76] F. STROCCHI, *Complex Coordinates and Quantum Mechanics*, Reviews of Modern Physics, 38 (1966), pp. 36–40.

[77] A. UHLMANN, *The metric of Bures and the geometric phase*, in Groups and Related Topics, Kluwer Academic Publisher, Dordrecht, (1992), pp. 267–274.

[78] R. F. WERNER, *Quantum states with Einstein-Podolsky-Rosen correlations admitting a hidden-variable model*, Phys. Rev. A, 40 (1989), pp. 4277–4281.

[79] H. WEYL, *Quantenmechanik und Gruppentheorie*, Zeitschrift für Physik, 46 (1927), pp. 1–46.

[80] ———, *The theory of groups and quantum mechanics*, Dover Books on Intermediate and Advanced Mathematics, Dover, New York, (1950).

[81] W. K. WOOTTERS, *Entanglement of Formation of an Arbitrary State of Two Qubits*, Phys. Rev. Lett., 80 (1998), pp. 2245–2248.

[82] P. ZANARDI, P. GIORDA, AND M. COZZINI, *Information-Theoretic Differential Geometry of Quantum Phase Transitions*, Physical Review Letters, 99 (2007), p. 100603.

I want morebooks!

Buy your books fast and straightforward online - at one of world's fastest growing online book stores! Environmentally sound due to Print-on-Demand technologies.

Buy your books online at
www.morebooks.shop

Kaufen Sie Ihre Bücher schnell und unkompliziert online – auf einer der am schnellsten wachsenden Buchhandelsplattformen weltweit! Dank Print-On-Demand umwelt- und ressourcenschonend produziert.

Bücher schneller online kaufen
www.morebooks.shop

KS OmniScriptum Publishing
Brivibas gatve 197
LV-1039 Riga, Latvia
Telefax: +371 686 204 55

info@omniscriptum.com
www.omniscriptum.com

Printed by Books on Demand GmbH, Norderstedt / Germany